INSECT
CELL
CULTURE
ENGINEERING

BIOPROCESS TECHNOLOGY

Series Editor

W. Courtney McGregor

Xoma Corporation
Berkeley, California

ADDITIONAL VOLUMES IN PREPARATION

Recombinant Microbes for Industrial and Agricultural Applications, *edited by Yoshikatsu Murooka*

Protein Purification Process Engineering, *edited by Roger Harrison*

INSECT CELL CULTURE ENGINEERING

edited by

Mattheus F. A. Goosen
Andrew J. Daugulis

Department of Chemical Engineering
Queen's University
Kingston, Ontario, Canada

Peter Faulkner

Department of Microbiology and Immunology
Queen's University
Kingston, Ontario, Canada

Marcel Dekker, Inc.   New York • Basel • Hong Kong

Library of Congress Cataloging-in-Publication Data

Insect cell culture engineering / edited by Mattheus F. A. Goosen,
   Andrew J. Daugulis, Peter Faulkner.
      p.   cm. — (Bioprocess technology ; v. 17)
   Includes bibliographical references and index.
   ISBN 0-8247-8944-X
   1. Animal cell biotechnology. 2. Insects—Viruses. 3. Insects—
Cytology. I. Goosen, Mattheus F. A. II. Daugulis, Andrew J.
III. Faulkner, Peter. IV. Series.
TP248.27.A53I57   1993
660'.6—dc20   92-44694
                                                           CIP

The publisher offers discounts on this book when ordered in bulk quantities.
For more information, write to Special Sales/Professional Marketing at the
address below.

This book is printed on acid-free paper.

MARCEL DEKKER, INC.
270 Madison Avenue, New York, New York 10016

Current printing (last digit):
10 9 8 7 6 5 4 3 2 1

PRINTED IN THE UNITED STATES OF AMERICA

Series Introduction

Bioprocess technology encompasses all the basic and applied sciences as well as the engineering required to fully exploit living systems and bring their products to the marketplace. The technology that develops is eventually expressed in various methodologies and types of equipment and instruments built up along a bioprocess stream. Typically in commercial production, the stream begins at the bioreactor, which can be a classical fermentor, a cell culture perfusion system, or an enzyme bioreactor. Then comes separation of the product from the living systems and/or their components followed by an appropriate number of purification steps. The stream ends with bioproduct finishing, formulation, and packaging. A given bioprocess stream may have some tributaries or outlets and may be overlaid with a variety of monitoring devices and control systems. As with any stream, it will both shape and be shaped with time. Documenting the evolutionary shaping of bioprocess technology is the purpose of this series.

Now that several products from recombinant DNA and cell fusion techniques are on the market, the new era of bioprocess technology is well established and validated. Books of this series represent developments in various segments of bioprocessing that have paralleled progress in the life sciences. For obvious proprietary reasons, some developments in industry, although validated, may be published only later, if at all. Therefore, our continuing series will follow the growth of this field as it is available from both academia and industry.

W. Courtney McGregor

Preface

Genetic engineering research and the biopharmaceutical industry are being revolutionized by developments in insect cell culture/baculovirus technology. This system has the ability to efficiently produce recombinant proteins with a structure and biological activity that closely resembles the native protein. Insect cells act as hosts for the baculoviruses, which are excellent vectors for genetic engineering because of their high expression rate and posttranslational processing capabilities. Insect cell culture may also be employed for producing insect pathogenic viruses for agricultural and forest pest control. As insecticides the viruses are pathogenic only to the target insects and are not hazardous to the environment. Recombinant baculoviruses may spare the use of chemical pesticides in biorational control programs.

The effective and economical cultivation of insect cells on a large scale is a key step toward commercial production of biologicals in insect cell culture. There is a need to consolidate and expand our fundamental understanding of virology and insect cell cultivation, the interactions between bioreactor performance and cell processes, and the protein chemistry of recombinant products. This book reviews the development of insect cell culture techniques for the production of recombinant proteins and insect pathogenic viruses, with consideration of process scale-up and reactor design. The problems associated with insect cell immobilization, the use of serum-free culture media, insect cell sensitivity to sparging aeration, and recent research into foreign gene expression in insect cells are also addressed. This book fills the void between laboratory research and pilot plant scale insect cell culture/baculovirus technology.

As its title suggests, the book covers the needs of industrial personnel involved in bioprocessing as well as biochemical engineering students, biochemists, and microbiologists and biologists who wish to specialize in animal cell culture. Since the background of the readers may vary, a special introductory chapter is included, as well as review chapters on insect pathogenic viruses, the molecular biology of baculoviruses, and bioreactor design. Following this, insect cell immobilization, applications in pharmaceutical research, scale-up, serum-free media, and foreign gene expression are covered. The book ends with a section on the future of insect cell culture engineering.

Finally, special thanks go to Dr. Brian Bellhouse of the Medical Engineering Unit at the University of Oxford for providing office space and a collegial atmosphere during Dr. Goosen's sabbatical in England, and to Mrs. Wendy Claye and Mrs. Margaret Martin for typing several of the manuscripts and handling the correspondence during a very hectic but enjoyable year.

Mattheus F. A. Goosen
Andrew J. Daugulis
Peter Faulkner

Contents

Contributors

Spiros N. Agathos   *Rutgers University, Piscataway, New Jersey*

Andrew J. Daugulis   *Queen's University, Kingston, Ontario, Canada*

Kees D. de Gooijer   *Wageningen Agricultural University, Wageningen, The Netherlands*

Peter Faulkner   *Queen's University, Kingston, Ontario, Canada*

Glenn P. Godwin   *GIBCO BRL Cell Culture R & D, Grand Island, New York*

Mattheus F.A. Goosen   *Queen's University, Kingston, Ontario, Canada*

Stephen F. Gorfien   *GIBCO BRL Cell Culture R & D, Grand Island, New York*

Donald L. Jarvis   *Texas A&M University, College Station, Texas*

John Kuzio   *Queen's University, Kingston, Ontario, Canada*

Janusz J. Malinowski   *Polish Academy of Sciences, Gliwice, Poland*

Annette L. Meyer   *Parke-Davis, Ann Arbor, Michigan*

Leonard E. Post   *Parke-Davis, Ann Arbor, Michigan*

Darrell R. Thomsen   *The Upjohn Company, Kalamazoo, Michigan*

Johannes Tramper   *Wageningen Agricultural University, Wageningen, The Netherlands*

Just M. Vlak   *Wageningen Agricultural University, Wageningen, The Netherlands*

Stefan A. Weiss*   *GIBCO BRL Cell Culture R & D, Grand Island, New York*

William G. Whitford   *GIBCO BRL Cell Culture R & D, Grand Island, New York*

**Current affiliation:* Calypte Biomedical, Berkeley, California

INSECT
CELL
CULTURE
ENGINEERING

1

Insect Cell Culture Engineering: An Overview

Mattheus F. A. Goosen   *Queen's University, Kingston, Ontario, Canada*

1.1 INTRODUCTION

With the discovery of genetic engineering in the 1970s the capabilities of the biopharmaceutical industry vastly expanded. The main driving force behind this revolution was economics. A company now had the ability to mass produce a high value product at modest cost. This was accelerated, initially, by the production of the first human insulin using the *E. coli/* plasmid system. The past ten years, however, has seen a sober reassessment of the capability of genetically modified bacteria to produce commercially valuable proteins. The most crucial factor behind this reevaluation was probably the inability of the host cell to perform many of the posttranslational modifications required for the correct biological functioning of many animal proteins. It now appears that the majority of economically attractive animal proteins will have to be produced on an industrial scale using insect and mammalian cells.

The earlier interest in insect cell cultivation stemmed from the prospect of mass producing viral insecticides as an environmentally sound alternative to chemical pest control. The current excitement about insect cell cultivation, on the other hand, is due to the recent development of baculovirus vectors for high level expression of many foreign genes whose products are correctly processed posttranslationally in the insect host cell. The most visible application of this system is the development of the first candidate vaccine against the AIDS virus, HIV, now in clinical trials.

One of the major advantages of the insect cell/baculovirus expression system over bacterial and mammalian expression systems is the very high expression of recombinant proteins, which are in many cases, antigenically, immunogenically, and functionally similar to their native counterparts (Table 1). Furthermore, baculoviruses are not pathogenic to vertebrates or plants and do not employ transformed cells (or transforming elements) as in mammalian expression systems. Many of the protein modification, processing and transport systems that occur in higher (eukaryotic) cells are utilized by the insect cell/baculovirus system. These modifications may be essential for the proper biological function of a recombinant protein. Typical product yields normally vary from a low of 1–5 mg/L for human interferon to a high of 600 mg/L for β-galactosidase.

Efficient cultivation methods are a key factor in the commercial exploitation of insect cell systems. The culturing conditions for insect and mammalian cells are similar. Most of the culture techniques have evolved through the modification of mammalian cell culture procedures.

Table 1 Recombinant Protein Expression with Insect Cell/Baculovirus System

Gene	Protein size (MW kD)	Modifications	Expression (mg/L)
Human interferon	17/21	Glycosylated, signal peptide cleaved, secreted	1–5
Chloramphenicol acetyl transferase	27	---	100
β-galactosidase	110	---	600
Simian rotavirus	41	---	50
Human reovirus	14.5	---	1–5
Human interleukin	15.5	Signal peptide cleaved, secreted	10–20
Influenza A	85	---	1–5

Source: M. D. Summers and G. E. Smith. (1985). Genetic engineering of the genome *Autographa californica* nuclear polyhedrosis virus, *Genetically Altered Viruses in the Environment* (B. Fields, M. Martin and D. Kamely, eds.), Banbury Rept. 22, Cold Spring Harbor, New York.

The culture systems for animal cells, for example, may be divided into two types, those in which the cells are grown attached to the surface of the culture vessel or some other solid support, and those in which the cells are grown suspended in the liquid medium. The former is usually referred to as substrate- or anchorage-dependent culture and the latter suspension culture. The most promising approach for insect cells involves cultivation in suspension using bioreactor technology that is currently applied successfully to microbial and mammalian cell product manufacturing. There have been reports on suspension cell growth in various types of vessels such as spinner flasks, stirred tanks and airlift bioreactors (Table 2). Aside from suspension culture, insect cells may also be immobilized in a polymer matrix, adsorbed unto surfaces, or entrapped behind a polymeric membrane. While this is being done to enhance cell densities and to aid downstream processing, this technology entails overcoming a slightly different set of problems (e.g., biocompatibility). Due to the high oxygen demand of production systems employing insect cells in suspension, mass transfer assisting operations such as agitation and air sparging need to be employed. This results in hydrodynamic shear stress, causing cell damage and death, which has been considered the key bottleneck in insect cell culture scale-up. The problem is aggravated by our limited understanding and sparse published literature on the behavior of insect cells growing in reactors.

1.2 THE PROBLEM OF SPARGING AND INSECT CELL CULTURE SCALE-UP

A variety of studies have been published on the sensitivity of insect cell suspension culture to scale-up (for reviews see Agathos et al., 1990; Wu et al., 1989; Shuler et al., 1990; Tramper et al., 1990; Goosen, 1992; Murhammer and Goochee, 1991). Caron et al. (1990), for example, assessed several types of stirred bioreactors for high level expression of a recombinant baculovirus, for the VP6 bovine rotavirus capsid protein, using Sf-9 insect cells in suspension culture. Scaling up to a surface aerated 4 liter marine impeller stirred tank bioreactor, cells densities of 5 $\times$ 10^6 viable cells/ml were routinely achieved, at 100 rpm, in medium (TNMFH) supplemented with 10% serum (fetal bovine serum) and 0.1% Pluronic F-68. However, growth was not as successful in the 4 liter bioreactor at 100 rpm when using serum-free medium (IPL/41). The sensitivity of the insect cell cultivation process is shown by the fact that when the concentration of Pluronic F-68 was increased to 0.3% and the stir speed was decreased 75 rpm, similar cell densities were obtained as

Table 2 Cultivation of Insect Cells in Suspension Culture: A Comparison of Culture Techniques

Cell line	Medium	Type of bioreactor	Critical shear stress N/m^2	Maximum cell density (cells/mL) × 10^{-6}	Comments	References
Aedes albopictus	Modified Eagle's 10% FBS	N/A	N/A	N/A		Spradling et al., 1975
Schneider Drosophila	Modified Eagle's 10% FBS	Spinner flask	N/A	10	Cells adapted to grow in suspension culture	Lengyl et al., 1975
TN-368	Modified TNM-Fh 10% FBS	Spinner flask (100 mL)	N/A	2 within 3–4 days	Cell growth enhanced by aeration	Hink and Strauss, 1976
		Bioflo (400 mL) MF-205, (2L)	N/A N/A	3 within 6–7 days	0.1% MC alleviated cell clumping	
		Vibromixer-glass	N/A	lower than other reactors used		
TN-368	TNM-FH, SC, 10% FBS	MF-205 (2-31) with marine impeller and air-sparging	1.5	1.75 in 72 h w/antifoam; 96 h w/o antifoam	0.02% antifoam added to avoid foaming, MC increased from 0.1 to 0.2% to lower shear	Hink and Strauss, 1980
Sf9	BML-TC/10, 10% FBS 0.1% MC	CSTR, 1L	1.5–3.0	3		Tramper et al., 1986
Sf9	Serum-free IPL-41 Basal with pluronic addition	airlift (internal loop)	N/A	5		Inlow et al., 1987

Cell line	Medium	Reactor		Growth	Comments	Reference
Sf-1PLB	Serum-free, with 0.5% egg yolk no antibiotics	Impeller agitated (10L)	N/A	2.4 w/o aeration; 4 w/aeration	Aeration by diffusion through a silicone tube system	Eberhard and Schugerl, 1987
Aedes albopictus	N/A	Spinner flask	N/A	5 at 100 rpm	0.5% CO_2 in air surface aeration	Agathos, 1988
		Jar reactor	1.0	N/A	0.1% MC added to medium enhanced cell growth	
Sf-9	EX-CELL	Spinner 8L	N/A	3.9 (6 days)	Cell viability decreased with scale-up; over 90% cell viability in 5L vessel while under 80% in 40L.	Weiss et al., 1988
		Airlift 5L	N/A	6.4 (8 days)		
		10L	N/A	3.0 (12 days)		
		40L	N/A	2.4 (8 days)		
IPLB-Sf-21	Grace's	shake-flask (125 mL)	N/A	3.2 (7 days)	0.2% pluronic F-68	Wu et al., 1990

Abbreviations: TN: Trichoplusia ni; MC: methylcellulose; N/A: not available; w/: with; w/o: without; Sf: *Spodoptera frugiperda*; FBS: Fetal bovine serum.
Source: Goosen, 1991.

in the serum supplemented culture. Some efforts were also directed at duplicating in sparged bioreactors, the growth curves obtained using surface aerated bioreactors. In one experiment in a 13 liter stirred bioreactor with serum-free medium containing 0.3% Pluronic F-68, culture was initiated with surface aeration and then switched to minimal pure oxygen sparging after the cells reached a density of 10^6/ml, without detrimental effects on cell viability. This is in agreement with recent reports (Maiorella et al., 1988; Murhammer and Goochee, 1988; Weiss et al., 1990), as well as work performed in our own laboratory with a 5 liter airlift reactor and serum-free medium (King et al., 1992) showing that sparging can be effectively used in insect cell culture in stirred and airlift bioreactors. In contrast, there have also been reports of serious problems in the scale-up of insect cell suspension cultures (Agathos et al., 1990; Wu et al., 1989; Shuler et al., 1990; Tramper et al., 1990; Tramper et al., 1986; Wu et al., 1990) (Table 3). Sparging and agitation appeared to be the key culprits in the loss of cell viability during culture. Let us therefore take a closer look at the mechanism of animal cell damage in bioreactors.

Cell damage due to excessive mechanical agitation may be associated with bubble entrainment in the fluid due to vortexing or cavitation (Kunas and Papoutsakis, 1990; Murhammer and Goochee, 1990). With insect cells in a 3 liter stirred bioreactor, Murhammer and Goochee found that moderate agitation at 300 rpm with a flat blade impeller, or at 500 rpm with a marine impeller, had no adverse effect on insect cell growth when there was no bubble entrainment via cavitation or vortexing. When bubble entrainment occurred at 850 rpm, cells died rapidly in the absence of Pluronic F-68, though they grew well in the presence of 0.2% Pluoronic F-68. We can speculate that the cell damage, which was reported by other groups to have been due to mechanical agitation, may actually have resulted from air sparging effects due to entrainment of air bubbles. Murhammer and Goochee also looked at different airlift reactors. They found that a high pressure drop across the gas sparger had a detrimental effect on cell growth, even in the presence of Pluronic F-68. They postulated that as the bubble detached from the orifice, cell damage was caused from turbulence and oscillations as fluid rushed into the region previously occupied by the bubble. In an interesting study, Kunas and Papoutsakis (1990) working with hybridoma cells in a surface aerated stirred reactor, showed that rapid movement of small bubbles within the medium (entrained bubbles) did not cause significant cell damage even at stir speeds up to 600 rpm. They further demonstrated that in the absence of a gas headspace (e.g., by using silicone tubing to aerate the medium) cells could grow even at stir speeds of up to 800 rpm. At higher stir speeds cell damage was observed. This was attributed to stresses from the turbulent liquid and related to the Kolmogorov eddy size. In the presence of a gas headspace, on the other hand, cell damage occurred at much lower stir speeds ($<$ 600 rpm). Their results suggested that this was probably due to bubble disengagement (bursting) at the air-liquid surface.

In sparged suspension culture, it has been proposed that Pluronic acts as a cell protective agent by stabilizing the foam layer on the liquid surface and thereby reducing film drainage and bubble bursting in the vicinity of cells (Handa et al., 1989). Bursting bubbles would cause nearby cells to rapidly oscillate, while the draining of liquid film between air bubbles would result in shear damage to any cells trapped between the bubbles. Protection may also arise from an interaction of surfactant and methyl cellulose polymers with the cell membrane, thus preventing direct cell/air interfacial contact (Goldblum et al., 1990, Murhammer and Goochee, 1990).

1.3 DEVELOPMENT OF SERUM-FREE MEDIUM

Due in part to the addition of fetal bovine or calf serum, typical insect cell culture media, such as Grace's, tend to be quite complex (Table 4). The presence of animal serum makes the medium expensive, thus imped-ing the economic scale-up of insect cell cultivation. Therefore, one of the major research thrusts, in industry as well as academia, has been aimed at developing serum-free insect cell culture media. Eliminating serum peptides will also enhance the downstream processing of recombi-nant products.

Over the past few years we have seen an increase in the use of serum-free medium for insect cell cultivation (Caron et al., 1990; Weiss et al., 1990; Lery and Fediere, 1990a; Lery and Fediere, 1990b). Serum-free cultures, however, may be more sensitive to agitation than serum supplemented cultures. Caron et al (1990) recommended increasing the total Pluronic F-68 concentration to 0.3% with IPL/41 serum-free me-dium. They observed a high cell growth rate but a decrease in recombi-nant protein production in the absence of serum. This differs from previ-ous results reported by Maiorella et al. (1988) and supports studies (unpublished) in our own laboratory that suggest that use of serum-free medium may sometimes result in significantly lower product concentra-tions. The high cell growth and comparatively low levels of recombinant protein production in serum-free medium points to major differences

Table 3 Comparison of Insect Cell Suspension Culture Methods

Type of bioreactor	Aeration	Impeller agitation	Medium serum supplemented	Pluronic F-68 (%)	Problems with cell growth	References
Stirred	Sparging	Yes	Yes	0	Yes	Agathos et al., 1990
	Surface	Yes	Yes	0	No	
Airlift	Sparging	No	Yes	0	Yes	
Stirred	Surface	Yes - 100 rpm	Yes	0.1	No	Caron et al., 1990
	Surface	100 rpm	No	0.1	Yes	
	Surface	75 rpm	No	0.3	No	
	Surface/Sparging	No	No	0.3	No	
Airlift	Sparging	No	No	0.1	No	Maiorella et al., 1988
Stirred	Sparging	Yes	Yes	0.2	No	Murhammer and Goocher, 1988
Airlift	Sparging	No	Yes	0.2	No	
Stirred	Surface/Sparging	Yes	No	---	No	Weiss et al., 1990
Stirred	Surface	Yes	Yes	0	Yes	Tramper et al., 1986c
Airlift	Sparging	No	Yes	0	Yes	

Airlift	Sparging	No	0.1	Yes	Wu et al., 1990
Stirred	Surface	Yes	0	Yes	Kunas and Papoutsakis,* 1990
	Silicone Tubing	< 600 rpm	0	No	
		> 600 rpm	0	Yes	
	(No entrainment)	800 rpm	0	No	
Stirred	Surface	Yes - 500 rpm	0	No	Murhammer and Goochee, 1990
		850 rpm	0	Yes	
			0.2	No	
Airlift	Sparging	No	0.2	Yes	
Bubble Column	Sparging	No	0.1	No	Handa et al., 1989
Stirred	Surface	Yes	0	No	Van Lier et al., 1990
Airlift	Sparging	No	0.1	No	King, et al. 1992

[a]Hybridoma cells
Source: Goosen, 1991b.

Table 4 Composition of Grace's Insect Cell Culture Medium (Grace, 1962)

Component	mg/L	Component	mg/L × 10^2
Inorganic Salts		*Vitamins*	
$CaCl_2$ (anhyd).	750	Biotin	1
KCl	4100	D-Ca pantothenate	2
$MgCl_2 \cdot 6H_2O$	2280	Choline chloride	20
$MgSO_4 \cdot 7H_2O$	2780	Folic acid	2
$NaHCO_3$	350	i-Inositol	2
$NaH_2PO_4 \cdot H_2O$	1013	Niacin	
		Para-aminobenzoic acid	2
Other Components		Pyridoxine HCl	2
α-Ketoglutaric acid	370	Riboflavin	2
Fructose	400	Thiamine HCl	2
Fumaric acid	55		
D-Glucose	700	*Serum Components*	
Malic acid	670	Bovine albumin fraction V	--
Succinic acid	60	Fetal bovine serum	--
Sucrose	26680	Whole egg ultrafiltrate	--

Amino Acids	
β-Alanine	200
L-Alanine	225
L-Arginine HCl	700
L-Asparagine	350
L-Aspartic acid	350
L-Cystine	22
L-Glutamic acid	600
L-Glutamine	600
Glycine	650
L-Histidine	2500
L-Isoleucine	50
L-Leucine	75
L-Lysine HCl	625
L-Methionine	50
L-Phenylalanine	150
L-Proline	350
DL-Serine	1100
L-Threonine	175
L-Tryptophan	100
L-Tyrosine	50
L-Valine	100

between factors promoting cell division and those involved in the late phase of infection with baculoviruses when recombinant proteins are produced. Serum-free medium formulations may need further modifications to allow for expression of more recombinants.

1.4 LIMITATIONS OF THE INSECT CELL/BACULOVIRUS SYSTEM

The insect cell/baculovirus expression system has its limitations. The expression of a foreign gene comes at the end of the infection cycle, at a time when the host cell is in the process of degradation. Bishop, in a recent review (1990), noted that it should not be surprising that processing of highly expressed recombinant products by the cell can be incomplete when very late baculovirus promoters are employed. Incomplete glycosylation, for example, was found for the glycoprotein of rabies (Prehaud et al., 1989). In another study Zu Putlitz et al. (1990) produced a monoclonal IgG antibody in baculovirus-infected insect cells. Both the light and the heavy chain cDNAs were introduced into the baculovirus in a single step. The apparent molecular weight of the recombinant heavy chain decreased after treatment with the enzyme endoglycosidase F/glycopeptidase F, indicating that the protein was in fact glycosylated. Of particular interest was the fact that Coomassie blue staining revealed a slight size difference between the glycosylated recombinant heavy chain and the control isolated from mouse ascites. This suggests a different glycosylation pattern (i.e., processed or cleaved high mannose type) compared to the native protein. The antibody molecule's antigen binding capability was not impaired.

Two successful large-scale studies were recently reported. A recombinant vaccine VP6 (Caron et al., 1990), produced in stirred tank bioreactors, was found to be fully effective as a subunit vaccine against bovine rotavirus. In another study recombinant EPO, produced in serum-free medium in stirred tanks, was analyzed by western blot and a specific bioassay and found to be fully glycosylated and resembled native protein (Weiss et al., 1990).

Another apparent limitation is that after prolonged passage through multiple infection cycles the ability of the virus to infect insect cells diminishes (i.e., the passage effect) (Tramper et al., 1990; Van Lier et al., 1990). Mutant virus particles are produced that interfere with the formation of regular virus. These so-called defective interfering particles (DIP) (K. De Gooijer, personal communication, Wickham et al., 1991), need the presence of wild type virus as a helper for multiplication. The passage effect limits the useful productivity time of a continuous system to about one month. Van Lier et al., studying the continuous production of baculovirus by Sf cells in a cascade of insect cell reactors, concluded that increasing the number of vessels accelerated the passage effect. These studies suggest that batch culture might be the preferred method of recombinant protein production.

How can we best develop a continuous insect cell culture system? Let us not forget that insect cell systems suitable for large-scale applications are not limited to those that serve as hosts to baculoviruses (i.e., the lepidopteran species) but also include cells of dipteran insects (mosquitoes) that transmit important diseases. In contrast to the lepidopteran cell/baculovirus vector system which leads to the lysis and eventual death of the host cells, dipteran cells appear to be prime candidates for stable long-term heterologous gene product formation (Agathos et al., 1990). This makes the latter more amenable to continuous culture production.

1.5 NEW DEVELOPMENTS

Mammalian cell suspension culture has been limited, to some extent, by relatively low cell densities and product concentrations (i.e., low productivities). The same problems may also be encountered in insect cell cultivation. Inlow et al., (1987), for example, reported maximum insect cell densities of 5.5×10^8 cells/ml in a spinner flask. Hink (1982), working with *T.ni* cells, obtained maximum polyhedra (AcNPV) concentrations of about 2.2×10^6 polyhedra per ml of medium at a cell density of 3.8×10^6 cells/ml of medium (i.e., about 60 polyhedra/cell). A major focus has, therefore, been placed on attempting to find cell culture methods that can increase the concentration of cells and cell products and permit cost-effective large-scale production. Several new methods of animal cell culture have been developed (mainly for mammalian cells) and involved the use of hollow fibers (Altshuler et al., 1986), gel entrapment (Scheirer et al., 1984), ceramic cartridges (Mosbach, 1984) and microcarriers (Dean et al., 1987).

Microencapsulation, a cell immobilization technique employed experimentally in an artificial pancreas (Sun and O'Shea, 1985) for the treatment of diabetes in animals, has been used industrially for enhanced production and recovery of high value biologicals from animal cells (Rupp, 1985). The capsule membrane selectively allows small molecules such as nutrients and oxygen to freely diffuse through, but prevents the passage of large molecules and cells. Posillico (1986) and Rupp (1985) reported the use of microencapsulation for the commercial production of monoclonal antibodies. Encapsulation of virus-infected cells by alginate can be used to produce cells in high density, although to get recombinant proteins, a temperature-sensitive vector is required so that the encapsulated (infected) cells can be induced to express a foreign protein when required (King et al., 1989). In addition, the complexity of the immobilization process and scale-up problems need to be addressed.

With regards to new developments in genetic engineering, recent studies by Jarvis (this text) have demonstrated the feasibility of utilizing a baculovirus immediate early gene promoter fused to either beta galactosidase or a human gene sequence to stably transform *Spodoptera frugiperda* cells. This has led to the emergence of the stable transformation approach, which can provide insect cell lines capable of expressing a foreign gene product continuously in the absence of a virus infection. This approach is currently being developed in both lepidopteran and dipteran insect cell systems. With further work in this area, it may be possible that the transformed insect cell approach will become as useful for foreign gene expression as the baculovirus approach. A valuable attribute to the baculovirus expression system is the ability to express more than one foreign gene at one time. Multiple gene vectors based on duplicated copies of AcNPV very late promoters have, for example, been prepared.

In the chapters to follow we will see reviews and updates of various aspects of insect cell culture technology. The material covered in this overview chapter will be expanded upon and added to.

REFERENCES

Agathos, S. N. (1988). "Insect Cell Cultivation in Bioreactors," Proceedings of the Engineering Foundation Conference on Cell Culture Engineering, Palm Coast, Florida, p. 10.

Agathos, S. N., Jeong, Y-H., Venkat, K. (1990). Growth kinetics of free and immobilized insect cell cultures (W. E. Goldstein, D. Dibiasio, H. Pedersen, eds.), *Ann. N. Y. Acad. Sci., 589* Biochem. Eng., pp. 372–398.

Altshuler, G. L., Dziewski, D. M., Sowek, J. A., and Belfort, G. (1986). Continuous hybridoma growth and monoclonal antibody production in hollow fiber reactors-separators, *Biotechnol. Bioeng., 28:* 646–658.

Bishop, D. L. (1990). Gene expression using insect cells and viruses: Current opinion, *Biotechnology, 1:* 62–67.

Caron, A. W., Archambault, J., and Massie, B. High-level recombinant protein production in bioreactors using the baculovirus-insect cell expression system, *Biotechnol. Bioeng., 36:* 1133–1140.

Croughan, M. S., Sayrre, E. S., Wang, D. I. C. Viscous reduction of turbulent damage in animal cell culture, *Biotechnol. Bioeng., 33:* 862–872.

Dean, R. C., Karkare, S. B., Ray, N. G., Runstadler, P. W., and Venkata-subramanian, K. (1987). Large-scale culture of hybridoma and mammalian cells in fluidized bed reactors, *Ann. N.Y. Acad. Sci., 506:* 129–146.

Eberhard, U., and Schügerl, K. (1987). Investigation of reactors for insect cell culture, *Develop Biol. Standard, 66:* 325–330.

Goldblum, S., Bae, Y. K., Hink, W. F., and Chalmers, J. (1990). Protective effect of methylcellulose and other polymers on insect cells subjected to laminal shear stress, *Biotechnol. Prog., 6:* 383–390.

Goosen, M. F. A. (1991). Insect cell cultivation techniques for the production of high-valued products, *Canadian J. of Chem. Engin. Biotechnol., 69, (2):* 450–456.

Goosen, M. F. A. (1992). Large-scale insect cell culture, *Current Opinion in Biotechnology, 3:* 99–104.

Grace, T. D. C. (1962). Establishment of four strains of cells from insect tissue grown *in vitro, Nature, 195:* 788.

Handa, A., Emery, A. N., and Spier, R. E. (1989). Effect of gas-liquid interfaces on the growth of suspended mammalian cells: Mechanism of cell damage by bubbles, *Enzyme Microb. Technol., 11* 230–235.

Hink, W. F., and Strauss, E. M. (1976). Growth of the Trichoplusia ni (TN-368) cell line in suspension culture, *Invertebrate Tissue Culture,* (E. Kurstak and K. Maramorosch, eds.), Academic Press, New York, pp. 297–300.

Hink, W. F., and Strauss, E. M. (1980). Semi-continuous culture of the TN-368 cell line in fermenters with virus production in harvested cells, *Invertebrate System in vitro,* (E. Kurstak, K. Maramorosch and A. Duebendorfer, eds.), Elsevier/North Holland Biomedical Press, Amsterdam, pp. 27–33.

Hink, W. F. (1982). Production of Autographa californica nuclear polyhedrosis virus in cells from large-scale suspension culture, *Microbial and Viral Pesticides,* (E. Kurstak, ed.), Marcel Dekker, New York, pp. 493–506.

Inlow, D., Harano, D., and Maiorella, B. (1987). "Large-Scale Insect Cell Culture for Recombinant Protein Production," Presented at American Chemical Society National Meeting, New Orleans, Louisiana.

King, G. A., Daugulis, A. J., Faulkner, P., Bayly, D., and Goosen, M. F. A. (1989). Alginate concentration a key factor in growth of temperature-sensitive baculovirus-infected insect cells in microcapsules, *Biotechnol. Bioeng., 34:* 1085–1091.

King, G. A., Daugulis, A. J., Faulkner, P., and Goosen, M. F. A. (1992). Recombinant β-galactosidase production in serum-free medium by insect cells in a 14-L airlift reactor, *Biotechnology Progress 8(b).*

Kunas, K. T., and Papoutsakis, E. T. (1990). Damage mechanisms of suspended animal cells in agitated bioreactors with and without bubble entrainment, *Biotechnol. Bioeng., 36:* 473–476.

Lengyel, J., Spradling, A., and Renman, S. (1975). Methods with insect cells in suspension culture III, *Drosophila malanogaster, Methods Cell Biol., 10:* 195–208.

Léry, X., and Fédiere, G. (1990a). Effect of different amino acids and vitamins of lepidopteran cell culture, *J. Invertebrate Pathol., 55:* 47–51.

Léry, X., and Fédiere, G. (1990b). A new serum-free medium for lepidopteran cell culture, *J. Invertebrate Pathol., 55:* 342–349.

Maiorella, B., Inlow, D., Shanger, A., and Harano, D. (1988). Large-Scale Insect Cell Culture for Recombinant Protein Production, *Biotechnology 6:* 1406–1410.

McQueen, A., Meilhoc, E., and Bailey, J. E. (19--). Flow effects on the viability and lysis of suspended mammalian cells, *Biotechnol. Lett., 9:* 832–836.

Mosbach, K. (1984). New immobilization techniques and examples of their application, *Ann. N.Y. Acad. Sci., 434:* 239–248.

Murhammer, D. W., and Goochee, C. F. (1988). Scale-up of insect cell cultures: Protective effects of Pluronic F-68, *Biotechnology 6:* 1411–1418.

Murhammer, D. W., and Goochee, C. F. (1990). Sparged animal cell bioreactors: Mechanism of cell damage and Pluronic F-68 protection, *Biotechnol. Prog., 6:* 391–397.

Murhammer, D. W., and Goochee, C. F. (1991). *Appl. Biochem. Biotechnol., 31:* 283–310.

Posillico, E. G. (1986). Microencapsulation technology for large-scale antibody production, *Bio/Technology,* 114–117.

Prehaud, C., Takehara, K., Flamand, A., and Bishop, D. H. L. (1989). Immunogenic and protective properties of rabies virus glycoprotein expressed by baculovirus vectors, *Virology, 173:* 390–399.

Rupp, R. G. (1985). Use of Cellular Microencapsulation in Large-Scale Production of Monoclonal Antibodies, *Large-scale Mammalian Cell Culture* (J. Feder and W. R. Tolbert, eds.), Academic Press, New York, pp. 19–38.

Scheirer, W., Nilsson, K., Merten, O. W., Katinger, H. W. D., and Mosbach, K. (1984). Entrapment of animal cells for the production of biomolecules such as monoclonal antibodies, *Dev. Biol. Stand., 55:* 155–161.

Shuler, M. L., Cho, T., Wickham, T., Ogonah, O., Kool, M., Hammer, D. A., Granados, R. R., and Wood, M A. (1990). Bioreactor development for production of viral pesticides or heterologous proteins in insect cell cultures, *Ann. N. Y. Acad. Sci. 589:* 399–421.

Spradling, A., Singer, R. H., Lengyel, J., and Penman, S. (1975). Methods with insect cells in suspension culture I: *Aedes albopictus, Methods Cell. Biol., 10:* 185–194.

Summers, M. D., and Smith, G. E. (1987). *A Manual of Methods for Baculovirus Vectors and Insect Cell Culture Procedures,* Texas Agricultural Station, Bulletin No. 1555, College Station, Texas.

Sun, A. M., and O'Shea, G. M. (1985). Microencapsulation of living cells—a long-term delivery system, *J. Controlled Release, 2:* 137–141.

Tramper, J., Williams, J. B., and Joustra, D. (1986). Shear sensitivity of insect cells in suspension, *Enzyme Microb. Technol. 8:* 33–36.

Tramper, J., and Vlak, J. M. (1986). Some engineering and economic aspects of continuous cultivation of insect cells for the production of baculoviruses. *Ann N.Y. Acad. Sci., 469:* 279–288.

Tramper, J., Van Den End, E. J., De Gooijer, C. D., Kompier, R., Van Lier, F. L. J., Usmany, M., and Vlak, J. M. (1990). Production of baculovirus in a continuous insect cell culture, *Ann. N. Y. Acad. Sci., 589:* 423–430.

Van Lier, F. L. J., Van Den End, E. J., De Gooijer, C. D., Vlak, J. M., and Tramper, J. (1990). Continuous production of baculovirus in a cascade of insect-cell reactors, *Appl. Microbiol. Biotechnol., 33:* 43–47.

Weiss, S. A., Belisle, B. W., DeGiovanni, A. M., Godwin, G. P., Kohler, J. P., and Summer, M D. (1988). Insect cells as substrates for biologicals, *Develop. Biol. Standard, 70:* 271–279.

Weiss, S. A., Gorfiens, S., Fike, R., Disorbo, D., and Jayme, D. (1990). "Large-Scale Production of Proteins Using Serum-Free Insect Cell Culture". Proceedings of the Ninth Australian Biotechnology Conference, Queensland, Australia, pp. 220–231.

Wickham, T. J., Davis, T., Granados, R. R., Hammer, D. A., Shuler, M. L., and Wood, H. A. (1991). Baculovirus defective interfering particles are responsible for variations in recombinant protein production as a function of multiplicity of infection, *Biotechnol. Lett., 13:* 483–488.

Wu, J., King, G., Daugulis, A. J., Faulkner, P., Bone, D. H., and Goosen, M. F. A. (1989). Engineering aspects of insect cell suspension culture: A review. *Appl. Microbiol Biotechnol., 32:* 249–255.

Wu, J., King, G., Daugulis, A. J., Faulkner, P., Bone, D. H., Goosen, M. F. A. (1990). Adaptation of insect cells to suspension culture, *J. Fermentation Bioeng., 70:* 90–93.

Zu Putlitz, J., Kubasek, W. L., Duchêne, M., Marget, M., Von Specht, B. U., and Domdey, H. (1990). Antibody production in baculovirus-infected insect cells. *Biotechnol., 8:* 651–654.

2

An Overview of the Molecular Biology and Applications of Baculoviruses

John Kuzio and Peter Faulkner   *Queen's University, Kingston, Ontario, Canada*

2.1 INTRODUCTION

Baculoviruses are almost exclusively viruses of insects. In the majority of the >600 baculovirus infections described, transmission occurs when insects ingest occlusion bodies in which infectious virus particles are embedded. Within insects the virus undergoes a complex replication cycle involving primary infection of the midgut followed by systemic spread of the lethal infection to many other insect tissues where progeny occlusion bodies are developed. The infection cycle is reestablished when occlusion bodies from dead insects contaminate potential feeding surfaces.

Historical writings have described catastrophic outbreaks of polyhedrosis disease in regions of Asia and Europe where silkworm (*Bombyx mori, B. mori*) were reared on a large scale (Benz, 1986). We now recognize that many species of baculoviruses predominantly infect insects from the order Lepidoptera (moths and butterflies). They have also been isolated from Hymenoptera (bees and wasps), Diptera (flies), Coleoptera (beetles), and Trichoptera (caddis flies) and have been identified at low incidence in other insect orders (Bilimoria, 1986; Federici, 1986; Rohrmann, 1986b). In addition, baculoviruses have been isolated from the cheliceratean order Arachnida (spiders), and the crustacean order Decapoda (shrimp and crabs). Baculoviruses have never been isolated from vertebrates or plants and their host range is considered restricted to the Arthropods.

2.2 TAXONOMY

The Baculovirus Family

The genus baculovirus (family Baculoviridae) is split into three subgenera for taxonomic purposes: group A, the nuclear polyhedrosis viruses (NPV); group B, the granulosis viruses (GV); and, group C, the nonoccluded viruses. Infection by NPV or GV is characterized by the production of paracrystalline, proteinaceous occlusion bodies (OB; also called polyhedra) in the nuclei of NPV, or cytoplasm of GV infected cells. The occlusion bodies of NPV are readily observed under the light microscope (Figure 1). Under the electron microscope they are seen to contain enveloped bundles of bacilliform virus nucleocapsids (Figure 2).

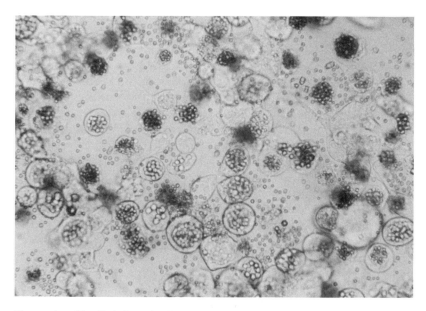

Figure 1. *Sf* cells infected with AcNPV. Light micrograph of *Sf* cells infected with AcNPV at 72 hr post infection. The cell nuclei are swollen and filled with occlusion bodies. Many of the cells have lysed and free occlusion bodies are seen in the surrounding media. (Magnification ×600.)

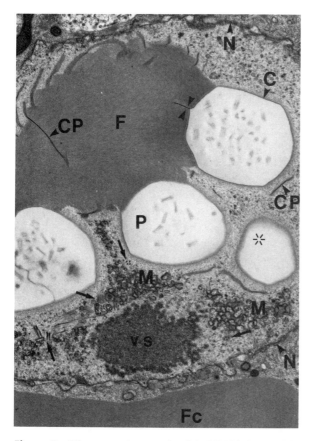

Figure 2 Electron micrograph of AcNPV infected *Sf* cell. Polyhedra (P) containing occluded virus are visible in the nucleus of an infected *Sf* cell at 36 hrs post infection. One occlusion body (*) is sectioned in a plane where no occluded virus is evident. The occlusion body calyx (C) is visible. Calyx precursors (CP) are present both in association with p10 fibrous bodies (F) and free in the nucleoplasm. A calyx precursor is seen attaching to an occlusion body (▶◀). Fibrous bodies are visible in both the nucleoplasm (F) and the cytoplasm (Fc). The nuclear membrane (N) is indicated. Short open-ended membrane profiles (M) are present near the nuclear periphery as are the remnants of the virogenic stroma (vs). Assembled nucleocapsids are seen in association with the membrane profiles in the process of being enveloped to form PDV (↑). Magnification ×18,440. (Photomicrograph courtesy of Greg V. Williams.)

The number of nucleocapsids per envelope is a morphological feature that is used to further divide group A into singly (SNPV) or multiply (MNPV) enveloped viruses. The availability of permissive cell lines such as those derived from *Spodoptera frugiperda* (*Sf;* fall army worm) and *Trichoplusia ni* (*T. ni;* cabbage looper) have made the NPV the most thoroughly researched of baculoviruses.

Baculovirus Species

Autographa californica multiply-enveloped Nuclear Polyhedrosis Virus (AcNPV; also written Ac*M*NPV) was isolated from the alfalfa looper and is the most extensively studied of the baculoviruses. It has been the subject of numerous biological and biochemical studies and has become a paradigm in baculovirus research. Whereas most NPV are restricted to infecting a single family or genus, AcNPV can infect at least 32 species from 12 families of Lepidopteran insects (Volkman and Keddie, 1990). The ability of AcNPV to replicate in several lines of cultured cells established the predominant use of this virus in early molecular studies of baculoviruses. Much of the work in deciphering the mechanisms of AcNPV infection, as well as the molecular biology of its host range and virulence, has been applicable to studies in other baculovirus species. For example, it is now evident that baculoviruses related to AcNPV, such as *Orgyia pseudotsugata* NPV (OpNPV; Douglas fir tussock moth), *Choristoneura fumiferana* NPV (CfNPV; spruce budworm), *Lymantria dispar* NPV (LdNPV; gypsy moth), or *Bombyx mori* NPV (BmNPV; silkworm) have much more similarity in their genomic organization than was initially suspected (Cochran et al., 1986; Blissard and Rohrmann, 1990; Rohrmann, 1986b).

2.3 BACULOVIRUS REPLICATION

In nature, NPV infection begins when an insect feeds on material contaminated with occlusion bodies. The occlusion bodies dissolve in the alkali conditions of the midgut (pH 9.5–11.5) and release PDV (polyhedron derived virus) which then infect columnar epithelial cells of the insect midgut. At least two factors have been identified as having supporting roles in the dissolution process. Alkaline proteases, associated with occlusion bodies purified from dead insects, may aid in the dissolution of occlusion bodies. These proteases are probably endogenous enzymes of the insect host gut (Rubenstein and Polson, 1983; Nagata and Tanada, 1983). Penetration of the peritrophic membrane (a secreted structure composed primarily of chitin and protein) by the PDV may be enhanced by a proteinaceous viral enhancing factor (Derksen and Granados, 1988).

PDV enter the epithelial cells by fusion of their envelope with the plasma membrane of the microvilli (Kawanishi et al., 1972; Granados, 1978; Granados and Williams, 1986) (Figure 3). Neutralization of infection by anti-PDV serum has been reported (Keddie and Volkman, 1985), implying that the process of PDV attachment and entry may be receptor mediated.

After membrane fusion, virus nucleocapsids transverse the cytoplasm to the nuclear membrane where they probably gain entry to the nucleus through nuclear pores (Granados and Lawler, 1981; Federici, 1986). Subsequently, viral nucleocapsids are uncoated in the nucleoplasm and the genomic DNA is exposed to the host cell transcription machinery. No viral proteins associated with the nucleocapsid are required for the initiation of transcription (mRNA production) or translation (protein synthesis) of early viral genes (Carstens et al., 1980; Kelly and Wang, 1981).

Ultrastructural and histochemical studies in cell culture have revealed that replication of viral DNA is associated with the appearance of a dense virogenic stroma within the nucleus and that virus nucleocapsids are assembled in the stroma (Federici, 1986; Granados and Williams, 1986). Initially the viral DNA may be associated with host nucleosomal proteins, however, as the infection progresses the host proteins are displaced by a virus encoded DNA binding protein (Wilson and Miller, 1986). Nucleocapsid assembly occurs by condensation of the virus genome to a tight helicoid nucleoprotein complex in a process that is mediated by the DNA binding protein. Hollow capsid shells have been seen aligned end-on the virogenic stroma and a subsequent maturation step results in the filling of the capsids with the nucleoprotein complex to produce nucleocapsids (Bassemir et al., 1983; Fraser, 1986).

In cell culture, at 24 hr post infection, maximal amounts of newly formed nucleocapsids are seen leaving the nucleus in the first stage of virus progeny development. This process appears to involve budding through the nuclear membrane, loss of the membrane during traversal of the cytoplasm, and final budding through a modified portion of the host cell plasma membrane. The plasma membrane is modified by incorporation of virus encoded proteins (Volkman, 1986). At this stage, the infectious particles are known as budded virus (BV). BV are released from the basolateral surface of the gut epithelial cells into the

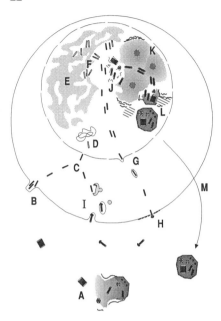

Figure 3 The baculovirus infectious cycle. Infection of an insect begins after the ingestion of a meal contaminated with polyhedra. Occlusion bodies dissolve in the midgut of the insect and release PDV (A). PDV fuse their envelopes with the cytoplasmic membranes of gut cells and the virions gain entry into the cytoplasm (B). After traversing the cytoplasm, virions enter the nucleus at a nuclear pore (C) and uncoat, releasing the viral genomic DNA into the cell nucleoplasm (D). A virogenic stroma develops and is the site for viral DNA replication and RNA transcription (E). Virus assembly occurs in the nucleus within the virogenic stroma (F). At approximately 24 hr, assembled virions leave the nucleus (G), traverse the cytoplasm, and BV bud through a modified cytoplasmic membrane (H). BV enter adjacent cells by adsorptive endocytosis (I) and initiate new rounds of infection (D,E,F). After 24 hr, a shift from BV to PDV production occurs. Assembled nucleocapsids are enveloped in the nucleus (J). Large amounts of polyhedrin are synthesized and transported into the nucleus where the protein begins to crystallize and occlude the enveloped virions. Sheet-like calyx precursors surround the nascent occlusion bodies (L). This process is believed to be mediated by p10 fibrous bodies (\\\). Mature occlusion bodies are liberated upon cell lysis and insect death (M).

haemocoel and a generalized infection of the insect ensues (Keddie et al., 1989).

The process of viral entry has been subjected to greater study for BV than for PDV and appears to be mechanistically different (Volkman and Keddie, 1990). Since BV are the form of the virus responsible for spread of the infection within the insect they have broad target organ specificity (Keddie et al., 1989). BV entry into cells by direct fusion at the plasma membrane, in the manner of PDV, may occur, albeit inefficiently. The principal mode of entry is receptor mediated adsorptive endocytosis (Volkman and Goldsmith, 1985). Once the virus is adsorbed onto the cell surface, it is internalized within clathrin coated vesicles. After uncoating of the vesicle (endosome) nucleocapsids gain entry into the host cell cytoplasm by fusion of the virus envelope with the cell membrane (Volkman, 1986). Primary lysosomes fusing with the virus containing endosomes and concomitant pH changes may trigger the membrane fusion process. After this step, the nucleocapsids may proceed with the infectious process as already described.

At later stages in the infection cycle, BV synthesis is abrogated, in both the insect and in cell culture, in favour of synthesis of the PDV phenotype. Large numbers of nucleocapsids associated with the virogenic stroma accumulate in the nucleus and become enveloped, either individually or in bundles. The origin of the PDV envelope is undefined. It may be synthesized *de novo* or possibly derived from a modified nuclear membrane (Stoltz et al., 1973; Mackinnon et al., 1974; Williams et al., 1989).

Also, late in the infection cycle, polyhedrin synthesis accelerates and the protein is transported into the nucleus where it begins to crystallize. In this process, PDV become surrounded by the polyhedrin matrix. Occlusion of the virus is dependent on it being enveloped since nonenveloped nucleocapsids have not been observed in occlusion bodies. Maturation of occlusion bodies is completed by their envelopment in a polyhedral calyx composed of sugars and at least one phosphoprotein (Whitt and Manning, 1988). Insect death due to polyhedrosis disease results in release of occlusion bodies and contamination of the substrate.

NPV infections are polyorganotropic and occlusion bodies are found in the nuclei of most larval tissues. After death and lysis of the host insect tissues, occlusion bodies protect PDV prior to ingestion by other larvae. Occlusion bodies play an important role in virus persistence in populations of insects that have seasonal feeding cycles (Jaques, 1977; Jaques, 1985).

2.4 BACULOVIRUS PHENOTYPES

Baculoviruses are unusual among animal viruses in that the nucleo-capsid of the virion lies within one of two forms of viral envelope (Volkman, 1986); thus each species of NPV exists in distinctive virion phenotypes (Figure 4): 1) the polyhedra derived virus (PDV; also

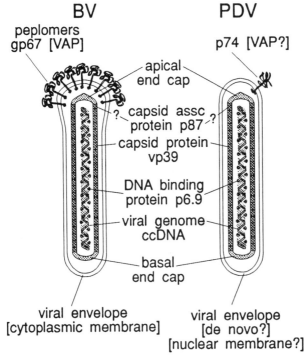

Figure 4 Baculovirus Virions. The ECV and PDV phenotypes of a baculovirus have a common structure, the nucleocapsid. Nucleocapsids are composed of a nucleoprotein core and a protein capsid shell. The nucleoprotein core contains the DNA of the viral genome as well as an abundant basic DNA binding protein, p6.9. The cylindrical portion of the capsid shell is composed of the protein vp39. In addition, the capsids have end cap structures that appear to have different ultrastructural appearances. One protein, p87, has been shown to be associated with capsids, however, its location in the structure has not yet been determined. The capsids are surrounded by lipid envelopes. The BV envelope is derived by budding through the plasma membrane. The PDV envelope may be synthesized de novo or may be derived from the host cell nuclear membrane. A BV enve-lope glycoprotein, gp67, has been shown by neutralization assays to be a viral attachment protein (VAP) and is thought to be the main component of the BV peplomers. In PDV, p74 is a candidate VAP.

called occluded virus, OV), the virus found in occlusion bodies, which makes natural horizontal transmission of the virus within an insect population possible; and 2) the budded virus (BV; also called extracellular virus, ECV; nonoccluded virus, NOV), which is the form of the virus responsible for systemic spread of infection within the insect and for transmission of the virus in cell culture. Within envelopes of both BV and PDV are rod-shaped nucleocapsids believed to be identical in biochemical and genetic composition (Cochran et al., 1986; Volkman, 1986). Thus the phenotypes differ only in the composition and origin of their envelopes.

Nucleocapsid Components

Baculovirus nucleocapsids comprise a nucleoprotein core and a protein shell (the capsid). The core contains the viral genome bound tightly to a basic DNA binding protein which functions to compact the large viral genome into a tight helicoid structure for efficient packaging. The primary sequence of the gene encoding the DNA binding protein has been determined for 3 baculovirus species, AcNPV (Wilson et al., 1987), OpNPV (Russell and Rohrmann, 1991) and BmNPV (Maeda et al., 1991a). The gene products are small basic peptides, rich in arginine residues (pI 12.8–12.9) that share considerable sequence identity. The proteins are also serine and threonine rich. Phosphorylation of serine and threonine residues, by a nucleocapsid associated protein kinase, may be instrumental in the virus uncoating process (Miller et al., 1983; Wilson and Consigli, 1985b; Wilson and Consigli, 1985a). Tyrosine residues are conserved and spaced throughout the length of the peptide. It has been proposed that the viral genome is condensed by ionic interactions between the arginine residues and the phosphopentose backbone of the DNA, while the tyrosine residues may intercalate among the stacked base pairs (Kelly et al., 1983). It has been noted that this protein bears a resemblance to some proteins in the protamine family (Tweeten et al., 1980; Rohrmann, 1991).

Baculovirus nucleocapsids are composed of a series of stacked rings rather than a helical structure common in other rod-shaped virions (Federici, 1986). The rings are in turn composed of 12 proteinaceous subunits. The major protein component of the capsid is encoded by the vp39 gene (Pearson et al., 1988; Thiem and Miller, 1989a; Blissard et al., 1989). The antisera to the capsid protein of OpNPV cross-reacts with a protein found in condensed chromosomes in uninfected cells (Pearson et al., 1988).

Ultrastructural studies revealed that the nucleocapsids display polarity in that they have distinctive caps at each end. These end caps have been described as the apical (pointed) end cap and the basal (rounded) end cap (Federici, 1986; Volkman and Keddie, 1990). The structures may reflect the mechanism of assembly of nucleocapsids. It is speculated that the apical end cap may mediate packing of the nucleoprotein into the capsid and may interact with membranes to trigger envelopment of nucleocapsid particles (Federici, 1986; Fraser, 1986). Proteins in the end caps may also have specific roles during the intracellular infectious process. Electron micrographs have shown an association of the nucleocapsid apical end caps and the nuclear pores (Granados and Lawler, 1981; Federici, 1986). A capsid associated protein, p87, has been identified as a component of BV, PDV, and purified capsids; it is not known if this protein lies within the cylindrical portion of the capsid or in an end cap (Muller et al., 1990).

Viral Envelopes

BV and PDV seem to differ only in the composition of their surrounding lipid envelopes and this is a reflection of their different routes of biosynthesis. The envelope lipids are considered to be derived from their host cell plasma membranes but the structural proteins embedded in them are unique to the baculovirus species. In the BV phenotype, gp67 (sometimes designated gp64), has been tentatively identified as the viral attachment protein that binds specifically to the cell receptor of a permissive host insect cell (Volkman, 1986). Monoclonal antibodies directed against gp67 neutralize BV, implicating this protein as the viral anti-receptor of BV (Volkman et al., 1984). The cellular receptor has not yet been identified. Ultrastructural features at the apical ends of BV, termed peplomers, are believed to be composed of multimeric gp67 (Volkman and Knudson, 1986; Volkman, 1986).

The gp67 protein has a characteristic amino terminal signal sequence and a carboxy terminal anchoring region that are common to transmembrane proteins (Whitford et al., 1989; Blissard and Rohrmann, 1989). The molecule may be stabilized in the membrane by covalently linked fatty acids (Roberts and Faulkner, 1989).

The PDV are the second virus phenotype found in infected cells. Studies in cell culture have established that PDV appear at a later stage in infection after BV morphogenesis has been attenuated. Since they are destined to become occluded in occlusion bodies, these particles do not bud from the cell; instead they accumulate in the cytoplasm (GV) or nucleus (NPV). The PDV capsids are also surrounded by a lipid envelope. The origin of this membrane has not been clearly established. It has been suggested that this membrane is synthesized de novo (Stoltz et al., 1973; Mackinnon et al., 1974), however, stepwise exhaustion (blebbing) of the nuclear membrane has been observed (Williams et al., 1989), suggesting that this host cell component could be a precursor for the PDV envelope. A candidate viral anti-receptor protein of PDV, p74, has been tentatively identified (Kuzio et al., 1989) (Kuzio and Faulkner, submitted manuscript).

2.5 POLYHEDRIN AND OTHER OCCLUSION BODY POLYPEPTIDES

Polyhedrin is the protein that forms the striking paracrystalline matrix of occlusion bodies (Figure 5), and has been the most extensively studied of baculovirus proteins (Rohrmann, 1986a). It accumulates in the later stages of virus infection, both in insects and in cell cultures, and crystallizes within nuclei producing large refractile bodies (1–15 μm) visible under the light microscope. In the process of crystallization, PDV become incorporated into the forming occlusion bodies. The high level of polyhedrin expression required to produce the numerous occlusion bodies found in infected cells has made studies on regulation of the polyhedrin gene preeminent in baculovirus research. Such knowledge has been applied to the development of recombinant baculoviruses to express a plethora of nonbaculovirus proteins in insect cells (Luckow and Summer, 1988; Luckow, 1991).

Occlusion bodies readily dissolve in alkaline solutions (i.e., $NaHCO_3$/ Na_2CO_3 pH ~ 10.5) to liberate PDV and yield polyhedral protein solutions. The polyhedrin gene is translated into a protein of approximately 29 kilodaltons. Polyhedrin genes from several NPV, as well as granulin genes of some GV, have been sequenced and found to share extensive amino acid identity (Rohrmann, 1986a). Synthesis of these proteins and formation of occlusion bodies is thought to provide a powerful selective advantage to baculoviruses by protecting the virions.

Mature occlusion bodies are surrounded by a carbohydrate rich structure called the occlusion body (or polyhedral) calyx (Whitt and Manning, 1988). This structure has also been named the polyhedral membrane or polyhedral envelope (PE), however, these terms are misleading because the calyx is not a lipid containing structure. In morphogenesis, the calyx is derived from electron dense sheets (also called spacers) that are present at sites of occlusion body formation. The calyx precursors have also been found within the p10 fibrous bodies that are associated with forming

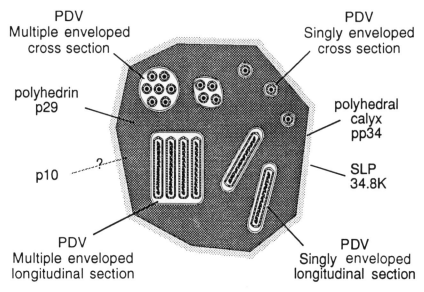

Figure 5 Baculovirus Occlusion Bodies. The occlusion body of the NPVs is a 1–15 µm polyhedral structure composed primarily of the protein polyhedrin. Embedded in it are the PDV particles. In the *M*NPVs the bundles are composed of 1 to 20 virions. The structure is surrounded by a calyx composed of carbohydrate and at least one phosphoprotein, pp34. A second protein, SLP (34.8K) has been detected on the occlusion body surface by cross-reacting antisera. The protein p10 has been reported to be present in occlusion bodies but its location and function is not yet known.

occlusion bodies (Mackinnon et al., 1974; van Lent et al., 1990; Russell and Rohrmann, 1990; Russell et al., 1991) (Figure 2). The calyx precursors are thought to be back to back associations of mature calyx structures (Mackinnon et al., 1974; Chung et al., 1980).

The calyx contains a thiol-linked phosphoprotein (Whitt and Manning, 1988). The sequences of the AcNPV (pp34) and OpNPV (p32) calyx protein genes have been determined (Oellig et al., 1987; Gombart et al., 1989). The amino and carboxy terminal regions of the encoded proteins show substantial similarity. The cysteine residues of the two proteins are within or near a variable hydrophilic central position of the peptides (Oellig et al., 1987; Gombart et al., 1989). Deletion of the gene encoding pp34 results in loss of the calyx precursors and in the formation of occlusion bodies lacking a calyx (Zuidema et al., 1989). Deletion of the p10 gene results in the loss of the p10 fibrous bodies and also generates occlusion bodies lacking a calyx (Williams et al., 1989). However, in this instance, calyx precursor structures are seen, but these fail to envelope occlusion bodies. Scanning electron micrographs of wild type and calyx negative occlusion bodies revealed that the latter were inordinately susceptible to mechanical disruption (Williams et al., 1989). Thus the role of the calyx appears to be to stabilize the occlusion bodies.

The p10 protein, like polyhedrin, is synthesized in large amounts late in the infection cycle. Immuno-electron microscopy has localized the protein to the fibrous bodies found in the cytoplasm and nucleus of infected cells (Van der Wilk et al., 1987; Russell et al., 1991). Recently p10 has been detected within polyhedra and associated with PDV (Russell et al., 1991). In addition, the protein has been found associated with cytoplasmic microtubules (Volkman and Zaal, 1990) and a cross-reaction between p10 antisera and uncharacterized cytoskeletal and nuclear components has been reported (Quant-Russell et al., 1987; Russell et al. 1991).

No fibrous bodies are seen in AcNPV p10 deletion mutants (AcNPV-p10) (Vlak et al., 1988; Williams et al., 1989). Cultured cells infected with AcNPV-p10 exhibit a normal virus growth cycle and produce BV and virulent occlusion bodies. AcNPV-p10 infected cells do not show the characteristic lytic cytopathic effect seen in wild type virus infections (Williams et al., 1989). Association of p10 with microtubules suggested that this protein could displace microtubule associated proteins (Quant-Russell et al., 1987), however, no differences were evident in patterns of microtubule reorganization in cells infected with wild type or AcNPV-p10 (Volkman and Zaal, 1990). To date, only one function has been assigned to p10: promoting the attachment of the polyhedral calyx to the occlusion bodies. The p10 genes of AcNPV and OpNPV show limited primary sequence identity (Kuzio et al., 1984; Leisy et al., 1986) implying that p10 may serve a structural rather than enzymatic role.

A second protein, with homology to the spheroidin protein of the *Choristoneura biennis* entomopoxvirus, has been found associated with occlusion bodies (Yuen et al., 1990; Vialard et al., 1990). Using cross-reacting anti-spheroidin antisera, the spheroidin-like protein (SLP) was localized by immunofluorescence microscopy to the polyhedral calyx. The previously sequenced AcNPV 34.8kD gene encodes SLP and it has been suggested that the 34.8kD gene encoding this protein was essential for virus replication in cell culture (Wu and Miller, 1989). However, it is difficult to reconcile this observation with the proposed role for the peptide as a structural protein of occlusion bodies.

A 25 kD protein has been implicated in a phenomenon associated with the yield of occlusion bodies in cultured cells (Beames and Summers, 1989). It had been observed that upon prolonged serial passage in cell culture, variants of the virus appear that produced fewer occlusion bodies per cell (FP variants) than wild type (MP variants) (Potter et al., 1976). This was subsequently shown to be the result of spontaneous transposon-mediated mutagenesis of the virus genome and resulted in the characterization of several transposons of insect cells (Blissard and Rohrmann, 1990). Analysis of one of the transposition hot spots on the virus genome revealed that strain conversion could be achieved by deletion of the 25 kD protein gene (Beames and Summers, 1989). Generation of FP phenotypes in culture has a negative implication for biotechnologists interested in maximizing yields of recombinant gene products.

2.6 MOLECULAR GENETICS OF BACULOVIRUSES

The Virus Genome

The genome of NPV is a covalently closed DNA molecule about 130,000 base pairs in length. Standard linear representations of the genetic map assume that the polyhedrin gene is located at the left end (Figure 6). The top strand of the DNA is called the R strand and corresponds to the sequence of clockwise transcripts on a circular map; the bottom (L strand) corresponds to counterclockwise transcripts (Vlak and Smith, 1982). Approximately 45% of the AcNPV genomic sequence has been published and/or deposited in GenBank.

Analysis of the sequenced regions of AcNPV reveals over 100 potential open reading frames (ORFs) that could encode peptides of at least 55 amino acids. Fortunately the task of analyzing the genome of AcNPV has been simplified. From the available sequence data, derived almost entirely from studies of AcNPV and OpNPV, it has become clear that genes of the baculoviruses are arranged as nonoverlapping, single exons with minimal regulatory regions between the ORFs. This observation eliminates ORFs that are embedded in other larger ORFs as potential coding regions. One case where this rule is violated is the immediate early gene IE1, which is transcribed by host cell RNA polymerase II and where splicing occurs to produce two related gene products IE0 and IE1 (Chisholm and Henner, 1988; Kovacs et al., 1991).

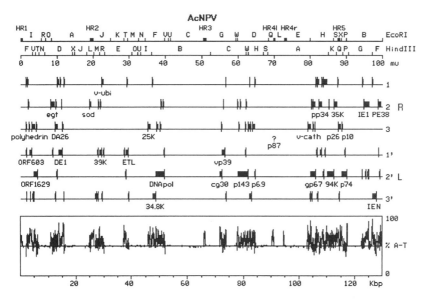

Figure 6 The Baculovirus Genome. A linear map of the genome of AcNPV in the standard orientation. The locations of cleavage sites for the restriction endonucleases EcoRI and HindIII are shown relative to their map position (map units, M.U.). All ORFs capable of encoding a protein of at least 55 amino acids are shown in each of three reading frames from the R (clockwise transcribed) and L (counterclockwise transcribed) strands of the genome. ORFs encoding an identified viral protein are labelled accordingly. The A-T (adenosine-thymidine) composition of the available viral genomic sequence is depicted graphically. The published sequences comprise 43% of the approximately 128 kilobase pairs (kbp) of the AcNPV genome.

No clustering of genes with respect to either function or temporal class has been identified in baculovirus genomes. Structural genes encoding virion proteins are intermixed with nonstructural genes, early genes with late genes. No strand preference is shown; the polyhedrin genes of AcNPV and OpNPV, though located in the same relative positions within their genomes, are transcribed from opposite strands (Leisy et al., 1984). Conservation of gene position is a common feature found in three well-studied species of NPV: AcNPV, OpNPV, and CfNPV (Leisy et al., 1984; Arif et al., 1985). The colinearity of the genome is presented as proof that these viruses have evolved from a common ancestor.

Another feature of the genome is the presence of interspersed repetitive DNA sequences, termed homologous regions (HRs) (Cochran and

Faulkner, 1983). HRs have been found in several baculoviruses though their copy number and position within the genome are not conserved between species (Cochran and Faulkner, 1983; Arif and Doerfler, 1984; Kuzio and Faulkner, 1984; McClintock and Dougherty, 1988). It has been demonstrated that HRs function as enhancer elements that may aid in the expression of viral genes early in infection (Guarino and Summers, 1986a).

Phases of Viral Protein Synthesis

As in other large DNA containing animal viruses, baculovirus protein synthesis is regulated at the level of gene transcription and temporal expression of groups of genes occurs in a cascade-like manner (Kelly, 1982; Friesen and Miller, 1986). In baculoviruses these groups of peptides are designated early (α and β), late (γ), and hyperexpressed late (δ).

Immediate early genes (α) are those that are transcribed by host RNA polymerase II very soon upon infection. Since NPV DNA is infectious, no viral proteins are required to initiate transcription of this set. Some immediate early genes have been characterized and sequenced: IE1 (Guarino and Summers, 1986; Theilmann and Stewart, 1991), IEN (Carson et al., 1991a), and PE38 (Krappa and Knebel-Morsdorf, 1991). IE1 and IEN have been identified as transactivators of β genes; they act to stimulate the expression of β genes and thus accelerate the infectious process (Guarino and Summers, 1986b; Carson et al., 1988). The AcNPV IE1 gene is currently the only known baculovirus gene to undergo splicing of its mRNA to produce two proteins, IE0 and IE1, which independently act to regulate both their own and β gene synthesis (Chisholm and Henner, 1988; Kovacs et al., 1991). IEN can transactivate IE1 as well as itself and thereby may play a role in accelerating IE1 synthesis (Carson et al., 1988; Carson et al., 1991a). Conversely, IE1 protein may down-regulate IEN synthesis (Carson et al., 1991b).

IEN and PE38 are transcribed from a bidirectional early promoter that is quasi-palindromic (Krappa and Knebel-Morsdorf, 1991). Transactivating capacity has not been ascribed to PE38. PE38 shares structural features with another AcNPV protein, CG30 (Thiem and Miller, 1989b); both have zinc fingers and leucine zippers, amino acid motifs that are characteristic of DNA binding proteins.

Baculovirus early promoters display no unique or unusual characteristics (Hoopes and Rohrmann, 1991). Early genes are transcribed by RNA polymerase II from eukaryotic-like promoters and mRNAs generally initiate at the consensus sequence CAGT (Blissard and Rohrmann, 1990).

HRs interspersed in the genome of AcNPV have been shown to have enhancer properties, mostly in conjunction with early gene expression (Guarino and Summers, 1986). Transcription of the AcNPV 39 kD gene was observed to be 1000-fold greater in the presence of a *cis*-linked AcNPV HR5 segment. HRs are composed of 2–8 copies of an imperfect, inverted repetitive sequence centered on EcoRI restriction sites (Guarino et al., 1986a). A single copy of the repeat is sufficient for enhancement of early gene expression (Guarino and Summers, 1986b). At least two protein factors have been identified in *Sf* cells that can form complexes with HRs (Guarino and Dong, 1991).

Delayed early genes (β) are transcribed in the presence of α gene products. A well-studied delayed early gene is the 39 kD gene. Its function is not known but antisera raised against a 39 kilodalton nuclear matrix-associated protein is reactive against the 39 kD gene product (Guarino and Smith, 1990; Wilson and Price, 1988). Several other β genes have been mapped using plasmid cotransfection expression assays with plasmids containing the IE1 gene (Guarino and Summers, 1987; Guarino and Summers, 1988). Delayed early genes may be expressed as immediate early genes but their expression is enhanced by α gene polypeptides (Guarino and Summers, 1987; Theilmann and Stewart, 1991).

Together, α and β protein synthesis constitute an early period of the infection cycle (0 to 6 hrs post infection), traditionally defined as that period up to the start of viral DNA synthesis. Late gene transcription coincides nominally with the onset of viral DNA synthesis at approximately 6 hrs post infection. At this point, transcription is believed to be directed by an as yet unidentified α-amanitin resistant RNA polymerase (Fuchs et al., 1983; Huh and Weaver, 1990).

NPV γ polypeptides include most virion structural proteins, and their synthesis extends through the period of BV and PDV assembly. At 12–18 hrs post infection, levels of host mRNA decline and there is a concomitant decrease in host protein synthesis (Ooi and Miller, 1988). A structural feature of baculovirus late genes is the presence of a common upstream promoter/transcription start sequence: ATAAG, termed the baculovirus late promoter (Blissard and Rohrmann, 1990).

An exact demarcation between α, β and γ gene synthesis is not possible due to temporal overlapping of gene transcription. Some early genes continue to be synthesized late in infection (Guarino and Summers, 1987). By contrast, gp67, formerly thought to be expressed exclusively as a late gene, can be detected as early as 2 hrs post infection; that is, prior to DNA synthesis (Blissard and Rohrmann, 1989).

The very late genes (δ) were thought to be expressed after τ gene transcription; however, δ genes, such as polyhedrin and p10, are now designated as hyperexpressed late genes. Whereas γ gene transcription is curtailed at late times in the infection cycle, δ genes continue to be transcribed until cell death. Hyperexpressed genes also initiate transcription at the same consensus baculovirus promoter sequence and are transcribed by an α-amanitin resistant RNA polymerase (Fuchs et al., 1983; Huh and Weaver, 1990; Blissard and Rohrmann, 1990). It is not known what viral and cellular factors regulate the expression of γ and δ class genes. *B. mori* cells infected with the heterologous baculovirus AcNPV express very low levels of polyhedrin (Summers et al., 1978; Luckow and Summers, 1988), an observation that suggests host cell factors may still contribute to δ gene expression.

The baculovirus genome is overtranscribed: many more species of viral mRNAs are produced than are thought required to express the potential number of genes in the genome (Friesen and Miller, 1986). Many of these RNAs are overlapping and have common 5' or 3' termini (Lubbert and Doerfler, 1984; Friesen and Miller, 1985; Rankin et al., 1986). In other cases portions of the genome are transcribed in both the sense and antisense directions (Guarino and Summers, 1987; Ooi and Miller, 1990; Ooi and Miller, 1991).

It has been demonstrated that polyhedrin mRNA synthesis can down-regulate the levels of a moderately expressed downstream antisense mRNA (Ooi and Miller, 1990). Two favored mechanisms for this process could be the disruption of the downstream promoter complexes by transcribing RNA polymerases from the polyhedrin promoter, or by the formation of RNA-RNA duplexes, a concomitant increase in RNA turnover and a disproportionate loss of the downstream mRNA (Ooi and Miller, 1990). Alternately, IE1 expression continues unabated late in infection despite the appearance of antisense late messages (Guarino and Summers, 1987).

Virus-Encoded Nonstructural Proteins

Viral structural proteins that become incorporated into virions and occlusion bodies have already been described. Other nonstructural viral proteins are synthesized in the course of infection. In some cases, such as the immediate early genes IE1 and IEN, gene function has been experimentally deduced (Guarino and Summers, 1987; Carson et al., 1991a).

In other cases, gene function can be inferred by identifying gene homologues in other organisms by primary amino acid sequence identity. Baculovirus genes have been sequenced that have no identifiable cellular homologues and no readily apparent function (Figure 6).

Sequence analysis has identified at least two nonstructural proteins involved in the replication of AcNPV DNA. A virus encoded DNA polymerase was identified by primary DNA sequence identity with other known DNA polymerases (Miller et al., 1981; Tomalski et al., 1988). The baculovirus DNA polymerase had been shown previously, through biological properties such as aphidicolin sensitivity, to be related to mammalian DNA polymerases α and δ and the DNA polymerases of herpesvirus, poxvirus, and adenovirus. A second protein, p143, has been identified as a putative helicase, a protein that unwinds DNA during the replication process (Lu and Carstens, 1991). Temperature sensitive mutants of p143 failed to replicate viral DNA at the nonpermissive temperature (Brown et al., 1979).

The AcNPV ETL gene has been shown to share sequence similarity with cyclin (Crawford and Miller, 1988; O'Reilly et al., 1989). AcNPV ETL deletion mutants displayed a 4–6 hr delay in expression of late proteins. Cyclins are a class of proteins involved in the transition of eukaryotic cells into the mitotic S phase (DNA replication) in cell division. Thus expression of a viral cyclin could advance host cells into S phase, thereby providing an environment more conducive to replication of viral DNA (O'Reilly et al., 1989).

Other baculovirus genes may have cellular homologues. Molting of insects is controlled by ecdysteroid hormones. An AcNPV gene, egt (UDP-glucuronosyl transferase), catalyses the transfer of glucose from UDP-glucose to this class of hormone and slows the molting of infected larvae (O'Reilly and Miller, 1989; O'Reilly and Miller, 1990). Fifth instar *Sf* larvae infected with mutant AcNPV lacking egt are not growth arrested and do not exhibit weight loss prior to death. Thus, infection with virus expressing egt results in larger larvae and more progeny occlusion bodies. Cellular glucuronosyl transferases are known to be membrane associated, but the viral protein is secreted. This difference is explained by the absence of 30 C terminal amino acids in egt required to anchor the protein to the cell membrane. It has been suggested that the ability to be secreted is a evolved in viral egt to allow it to act systemically.

Other baculovirus genes with cellular homologues recently identified are: a viral ubiquitin (v-ubi) (Guarino, 1990), and a viral superoxide dismutase (sod) (Tomalski et al., 1991) but their roles in the infectious process have yet to be identified. AcNPV sod deletion mutants do not show deficiencies in replication in cell culture and in insects. No mutants in the v-ubi gene have been isolated, suggesting that this gene may be essential for virus replication.

Recently, a gene with homology to the papain family of cysteine proteinases has been identified (J. Kuzio and P. Faulkner, submitted manuscript). This gene does not appear to be essential for virus infection of cell cultures or insects.

2.7 APPLICATIONS OF RECOMBINANT BACULOVIRUSES

Intracellular events in replication are considered similar in cultured cells and in insect tissues; however, in cell culture infection is initialized by inoculating cells with BV rather than PDV. Progeny BV produced in permissive cells transmit the virus to all other cells in culture. During the late period of infection occlusion bodies accumulate but these play no role in cell–cell transmission of the virus. This lack of requirement for the PDV phenotype in cell culture is the key factor in utilizing the polyhedrin gene promoter for expression of foreign gene products in insect cells.

Baculovirus Expression Vectors

The development of baculovirus transfer vectors and the expression of foreign protein from the polyhedrin promoter in insect cells infected with a recombinant AcNPV launched a new era in baculovirology (Smith et al., 1983; Miller, 1984; Pennock et al., 1984). The general aspects of baculovirus expression vector technology have been covered in several reviews (Summers and Smith, 1987; Miller, 1988; Luckow and Summers, 1988; Cameron et al., 1989; Possee et al., 1990; Luckow, 1991; Fraser, 1992), and general techniques for producing baculovirus vectors that express foreign genes are described in Chapter 5. Advantages and drawbacks of using baculovirus expression vectors in biotechnology are described in Chapters 8 and 9.

Recent developments in construction of baculovirus transfer vectors have endeavoured to make them more user friendly. Positive selection systems using β-gal (Vialard et al., 1990a; Vlak et al., 1990; Zuidema et al., 1990) or occlusion body positive (OB$^+$) screening (Weyer et al., 1990; Wang et al., 1991) have replaced the more cumbersome screening procedure based on selection of recombinant strains which are unable to produce occlusion bodies. Vectors containing alternate (Weyer et al., 1990;

Hill-Perkins and Possee, 1990) and modified (Ooi et al., 1989; Wang et al., 1991; Thiem and Miller, 1990) baculovirus gene promoters have been constructed in an attempt to improve levels of transcription. Optimization of the cloning site with respect to the vector leader sequences has been shown to result in enhanced translation of some recombinant proteins (Matsuura et al., 1987; Luckow and Summers, 1989; Page, 1989). Hormone stimulation of cultured cells has also increased yields of recombinant proteins (Sarvari et al., 1990). Filamentous bacteriophage origins of replication have been introduced into transfer vectors to facilitate modifying recombinant plasmids using simple oligonucleotide mutagenesis procedures (Livingstone and Jones, 1989; Hasemann and Capra, 1990). Techniques have been developed to enhance the frequency of recombination with the transfer vector to increase the ratio of recombinant to parental virus (Peakman et al., 1989; Mann and King, 1989; Kitts et al., 1990).

Established techniques to rear silkworm larvae have encouraged attempts to use the insects as living factories for the production of recombinant proteins from BmNPV expression vectors (Hasemann and Capra, 1990; zu Putlitz et al., 1990). *T. ni* larvae have also been used to express recombinant proteins using AcNPV-derived vectors (Medin et al., 1990; Price et al., 1989). It is not established whether the levels of protein produced in pilot scale experiments such as these can be translated into consistent large-scale production.

Foreign Gene Products Expressed in Insect Cells

The range of gene products expressed in insect cells extends from antibodies (Hasemann and Capra, 1990; zu Putlitz et al., 1990) to zinc finger proteins (Ollo and Maniatis, 1987). Single cistron, intronless genes are preferred. Limited studies have shown that splicing of transcripts containing introns will occur, leading to expression of low levels of recombinant proteins (Jeang et al., 1987; Lanford, 1988; Iatrou et al., 1989). Inefficient splicing of late mRNA containing introns expressed by recombinant baculoviruses may account for depressed levels of protein synthesis. In one study, only 25% of the recombinant mRNA was correctly spliced (Iatrou et al., 1989). Splicing efficiency was lower later in infection, implying that the cell splicing machinery may be disrupted at these times. Attempts to express multicistron mRNAs from a single δ promoter have shown that downstream genes are not expressed, which suggests reinitiation of translation does not occur on mRNA in infected cells (Hasemann and Capra, 1990).

Posttranslational modifications of proteins, including glycosylation, acylation, phosphorylation, signal peptide cleavage, secretion, and subunit assembly, have been observed for proteins expressed by baculovirus expression vectors (Luckow, 1991). Glycosylation has been the most extensively studied posttranslational process. Both *N*- and *O*-linked glycosylation of recombinant proteins has been observed. *N*-linked glycosylation is similar, but not identical, to the glycosylation seen in mammalian cells. Glycoproteins expressed in insect cells do not undergo the high mannose to complex *N*-linked oligosaccharide conversion; processing of the *N*-linked sugar side chains appears to be blocked at the trimming stage of the high mannose form of the side chains. Baculovirus vectors have been used to express immunologically reactive proteins as well as proteins with enzymatic activities such as DNA polymerases, protein kinases, and proteases. Lists of foreign genes expressed by baculovirus expression vectors have been published in other reviews (Luckow and Summers, 1988; Summers, 1989; Luckow, 1991).

Insect Pest Management and Baculovirus-Based Insecticides

Expanded use of baculoviruses in integrated biorational approaches to control insect pests will require an understanding of baculovirus pathogenesis, particularly with regard to viral and insect cell factors that affect host range and virulence. In their natural habitat, many baculoviruses are of low to moderate virulence. Thus an objective of researchers is to engineer strains with enhanced virulence and broader host range. Little progress has been made in these endeavors mainly due to a lack of understanding of the molecular basis of insect pathogenesis and identification of relevant genes and pathways. A more promising approach has been developing novel virus strains to arrest insect feeding as soon as possible after ingestion of virus, thus protecting plants from damage.

One approach to the problem has been to engineer baculoviruses that express insect-specific paralysing toxins. The δ endotoxin of *Bacillus thuringiensis* has an established role as a biological pesticide (Davidson, 1989). This gene has been inserted into a baculovirus and been expressed in an immunologically reactive and biologically active form. Recombinant viruses were isolated that directed the expression of biologically active protein in insect cells; however, preliminary studies showed no significant change in either the LD_{50} (mean number of deaths) or LT_{50} (mean time to death) when the virus was fed as occlusion bodies (Merryweather et al., 1990; Martens et al., 1990). By contrast, larvae ceased feeding and eventually died when presented with leaves contaminated with toxin-producing insect cells (Possee et al., 1991). It was demonstrated that recombinant viruses expressing δ endotoxin may have an impact on insect pest management by reducing feeding damage.

Insect-specific scorpion neurotoxins have been introduced into baculoviruses to attempt to generate paralytic strains. Synthesis of *Buthus eupeus* BeIT toxin was poor when a synthetic gene encoding this short peptide was expressed (Carbonell et al., 1989). A high levels of expression of BeIT toxin was detected when the gene was fused to the N terminal portion of polyhedrin; however, the protein lacked biological activity, possibly due to abnormal protein folding. A second recombinant expressed modest levels of BeIT when fused to the cleavable signal peptide from human β interferon, but this recombinant also lacked BeIT toxicity. Recently, scorpion toxin AaIT (*Androctonus australis*) was expressed using recombinant AcNPV and BmNPV. The toxin gene was fused to either the AcNPV gp67 signal peptide (Stewart et al., 1991) or the leader sequence from bombyxin (Maeda et al., 1991b; McCutchen et al., 1991). In both instances paralytic activity, a decrease in LT_{50}, and a reduction in feeding damage were observed in insects infected with the recombinant viruses. It has not been established whether the differences between BeIT and AaIT toxic effects in these recombinant viruses were due to the toxins or to signal peptides used.

A second insect specific toxin derived from a spider mite, *Pyemotes tritici*, has also been inserted into baculoviruses (Tomalski and Miller, 1991). Insects infected with recombinant viruses expressing the toxin became paralysed within 2 to 3 days post infection.

An alternate approach to increase virulence has been to engineer baculoviruses that express hormones that alter the physiology of the insect. A recombinant NPV has been created carrying a synthetic insect diuretic hormone (Maeda, 1989). Infection with this virus caused alterations in larval fluid metabolism and resulted in a decrease of the LT_{50} by 20% as compared to wild type virus. Infection of larvae was by injection of BV rather than by feeding occlusion bodies, and analysis of the true biological properties of this virus *in vivo* remains to be done. Promising results have also been obtained from AcNPV expressing juvenile hormone esterase (JHE) (Hammock et al., 1990; Posse et al., 1991). Feeding insect larvae maintain nominal titres of juvenile hormone (JH), but increased activity of JHE at the onset of metamorphosis results in a dramatic drop of JH and cessation of feeding behavior. First instar larvae infected with the JHE recombinant virus were stunted compared to larvae infected with the wild type baculovirus probably because feeding in the experimental group was reduced. However, experiments with later instar larvae showed no difference in the virulence of recombinant and wild type AcNPV. This was ascribed to low levels of synthesis of recombinant JHE in insects and *in vivo* inactivation of protein, resulting in insufficient enzyme activity in the larger larvae.

ACKNOWLEDGMENTS

We acknowledge financial support from the Medical Research Council of Canada and Insect Biotech Canada.

REFERENCES

Arif, B. M. and Doerfler, W. (1984). Identification and localization of reiterated sequences in the *Choristoneura fumiferana* MNPV genome, *Embo J.*, *3:* 525–529.

Arif, B. M., Tjia, S. T., and Doerfler, W. (1985). DNA homologies between the genomes of *Choristoneura fumiferana* and *Autographa californica* nuclear polyhedrosis viruses, *Virus Res., 2:* 85–94.

Bassemir, U., Miltenburger, H. G., and David, P. (1983). Morphogenesis of nuclear polyhedrosis virus from *Autographa californica* in a cell line from *Mamestra brassicae* (Cabbage moth). Further aspects of baculovirus assembly, *Cell Tissue Res., 228:* 587–595.

Beames, B. and Summers, M. D. (1989). Location and nucleotide sequence of the 25K protein missing from baculovirus few polyhedra (FP) mutants, *Virology, 168:* 344–353.

Benz, G. A. (1986). Introduction: Historical perspectives, *The Biology of Baculoviruses* (R. R. Granados and B. A. Federici, eds.), CRC Press, Boca Raton, Florida, pp. 1–35.

Bilimoria, S. L. (1986). Taxonomy and identification of baculoviruses, *The biology of baculoviruses* (R. R. Granados and B. A. Federici, eds.), CRC Press, Boca Raton, Florida, pp. 37–59.

Blissard, G. W., Quant-Russell, R. L., Rohrmann, G. F., and Beaudreau, G. S. (1989). Nucleotide sequence, transcriptional mapping, and temporal expression of the gene encoding p39 a major structural protein of the multicapsid nuclear polyhedrosis virus of *Orgyia pseudotsugata, Virology, 168:* 354–362.

Blissard, G. W. and Rohrmann, G. F. (1989). Location, sequence, transcriptional mapping and temporal expression of the gp64 gene of the *Orgyia pseudotsugata* multicapsid nuclear polyhedrosis virus, *Virology, 170:* 537–555.

Blissard, G. W. and Rohrmann, G. F. (1990). Baculovirus diversity and molecular biology, *Ann. Rev. Entomol., 35:* 127–155.

Brown, M., Crawford, A. M., and Faulkner, P. (1979). Genetic analysis of a baculovirus *Autographa californica* nuclear polyhedrosis virus. Isolation of temperature sensitive mutants and assortment into complementation group. *J. Virol., 31:* 335–387.

Cameron, I. R., Possee, R. D., and Bishop, D. H. L. (1989). Insect cell culture technology in baculovirus expression systems, *TIBTECH, 7:* 66–70.

Carbonell, L. F., Hodge, M. R., Tomalski, M. D., and Miller, L. K. (1989). Synthesis of a gene coding for an insect specific scorpion neurotoxin and attempts to express it using baculovirus vectors, *Gene, 73:* 409–418.

Carson, D. D., Guarino, L. A., and Summers, M. D. (1988). Functional mapping of an AcNPV immediate early gene which augments expression of the IE-1 transactivated 39K gene, *Virology, 162:* 444–451.

Carson, D. D., Summers, M. D., and Guarino, L. A. (1991a). Molecular analysis of a baculovirus regulatory gene, *Virology, 182:* 279–286.

Carson, D. D., Summers, M. D., and Guarino, L. A. (1991b). Transient expression of the *Autographa californica* nuclear polyhedrosis virus immediate-early gene, IE-N, is regulated by three viral elements, *J. Virol., 65:* 945–951.

Carstens, E. B., Tjia, S. T., and Doerfler, W. (1980). Infectious DNA from *Autographa californica* nuclear polyhedrosis virus, *Virology, 101:* 311–314.

Chisholm, G. E. and Henner, D. J. (1988). Multiple early transcripts and splicing of the *Autographa californica* nuclear polyhedrosis virus E-1 gene, *J. Virol., 62:* 3193–3200.

Chung, K. L., Brown, M., and Faulkner, P. (1980). Studies on the morphogenesis of polyhedral inclusion bodies of a baculovirus *Autographa californica* NPV, *J. Gen. Virol., 46:* 335–347.

Cochran, M. A. and Faulkner, P. (1983). Location of homologous DNA sequences interspersed at five regions in the baculovirus AcMNPV genome, *J. Virol., 45:* 961–970.

Cochran, M. A., Brown, S. E., and Knudson, D. L. (1986). Organization and expression of the baculovirus genome, *The Biology of Baculovirus* (R. R. Granados and B. A. Federici, eds.), CRC Press, Boca Raton, Florida.

Crawford, A. M. and Miller, L. K. (1988). Characterization of an early gene accelerating expression of late genes of the baculovirus *Autographa californica* nuclear polyhedrosis virus, *J. Virol., 62:* 2773–2781.

Davidson, E. W. (1989). Insect cell cultures as tools in the study of bacterial protein toxins, *Adv. Cell Cult., 7:* 125–146.

Derksen, A. C. G. and Granados, R. R. (1988). Alteration of a lepidopteran peritrophic membrane by baculoviruses and enhancement of viral infectivity, *Virology, 167:* 242–250.

Federici, B. A. (1986). Ultrastructure of Baculoviruses, *The Biology of Baculoviruses* (R. R. Granados and B. A. Federici, eds.), CRC Press, Boca Raton, Florida, pp. 61–88.

Francki, R. I. B., Fauquet, C. M., Knudson, D. L., and Brown, F. (1991). Classification and nomenclature of viruses, *Arch. Virol.,* Supplementum 2.

Fraser, M. J. (1986). Ultrastructural observations of virion maturation in *Autographa californica* nuclear polyhedrosis virus infected *Spodoptera frugiperda* cell cultures, *J. Ultrastr. Mol. Res., 95:* 189–195.

Fraser, M. J. (1992). The baculovirus infected insect cell as a eukaryotic gene expression system, *Curr. Top. Microbiol. Immunol., 158:* 131–172.

Friesen, P. D. and Miller, L. K. (1985). Temporal regulation of baculovirus RNA: Overlapping early and late transcripts, *J. Virol., 54:* 392–400.

Friesen, P. D. and Miller, L. K. (1986). The regulation of baculovirus gene expression, *Curr. Top. Microbiol. Immunol., 131:* 31–49.

Fuchs, L. Y., Woods, M. S., and Weaver, R. F. (1983). Viral transcription during *Autographa californica* nuclear polyhedrosis virus infection: a novel RNA polymerase induced in infected *Spodoptera frugiperda* cells, *J. Virol., 48:* 641–646.

Gombart, A. F., Pearson, M. N., Rohrmann, G. F., and Beaudreau, G. S. (1989). A baculovirus polyhedral envelope-associated protein: Genetic location, nucleotide sequence, and immunocytochemical characterization, *Virology, 169:* 182–193.

Granados, R. R. (1978). Early events in the infection of *Heliothis zea* midgut cells by a baculovirus, *Virology, 90:* 170–174.

Granados, R. R. and Lawler, K. A. (1981). *In vivo* pathway of *Autographa californica* baculovirus invasion and infection, *Virology, 108:* 297–308.

Granados, R. R. and Williams, K. A. (1986). *In vivo* infection and replication of baculoviruses, *The Biology of Baculoviruses* (R. R. Granados and B. A. Federici, eds.), CRC Press, Boca Raton, Florida, pp. 89–108.

Guarino, L. A. and Summers, M. D. (1986a). Interspersed homologous DNA of *Autographa californica* nuclear polyhedrosis virus enhances delayed early gene expression, *J. Virol., 60:* 215–223.

Guarino, L. A. and Summers, M. D. (1986b). Functional mapping of a transactivating gene required for expression of a baculovirus delayed early gene, *J. Virol., 57:* 563–571.

Guarino, L. A., Gonzalez, M. A., and Summers, M. D. (1986). Complete sequence and enhancer function of the homologous DNA regions of *Autographa californica* nuclear polyhedrosis virus, *J. Virol., 60:* 224–229.

Guarino, L. A. and Summers, M. D. (1987). Nucleotide sequence and temporal expression of a baculovirus regulatory gene, *J. Virol., 61:* 2091–2099.

Guarino, L. A. and Summers, M. D. (1988). Functional mapping of *Autographa californica* nuclear polyhedrosis virus genes required for late protein expression, *J. Virol., 62:* 463–471.

Guarino, L. A. (1990). Identification of a viral gene encoding a ubiquitin like protein, *Proc. Natl. Acad. Sci. USA, 87:* 409–413.

Guarino, L. A. and Smith, M. W. (1990). Nucleotide sequence and characterization of the 39K gene region of *Autographa californica* nuclear polyhedrosis virus, *Virology, 179:* 1–8.

Guarino, L. A. and Dong, W. (1991). Expression of an enhancer-binding protein in insect cells transfected with the *Autographa californica* nuclear polyhedrosis virus IE1 gene, *J. Virol., 65:* 3676–3680.

Hammock, B. D., Bonning, B. C., Possee, R. D., Hanzlik, T. N., and Maeda, S. (1990). Expression and effects of the juvenile hormone esterase in a baculovirus vector, *Nature, 344:* 458–461.

Hasemann, C. A. and Capra, J. D. (1990). High-level production of a functional immunoglobulin heterodimer in a baculovirus expression system, *Proc. Natl. Acad. Sci. USA, 87:* 3942–3946.

Hill-Perkins, M. S. and Possee, R. D. (1990). A baculovirus expression vector derived from the basic protein promoter of the *Autographa californica* nuclear polyhedrosis virus, *J. Gen. Virol., 71:* 971–976.

Hoopes, R. R. J. and Rohrmann, G. F. (1991). In vitro transcription of baculovirus immediate early genes. Accurate mRNA initiation by nuclear extracts from both insect and human cells, *Proc. Natl. Acad. Sci. USA, 88:* 4513–4517.

Huh, N. E. and Weaver, R. F. (1990). Identifying the RNA polymerases that synthesize specific transcripts of the *Autographa californica* nuclear polyhedrosis virus, *J. Gen. Virol., 71:* 195–201.

Iatrou, K., Meidinger, R. G., and Goldsmith, M. R. (1989). Recombinant baculoviruses as vectors for identifying proteins encoded by intron-containing members of complex multigene families, *Proc. Natl. Acad. Sci. USA, 86:* 9129–9133.

Jaques, R. P. (1977). Stability of entomopathogenic viruses, *J. Entomol. Soc. Am., 10:* 99–116.

Jaques, R. P. (1985). Stability of insect viruses in the environment, *Viral Insecticides for Biological Control* (K. Maramorosch and K. E. Sherman, eds.), Academic Press, New York, pp. 285–360.

Jeang, K.-T., Holmgren-Konig, M., and Khoury, G. (1987). A baculovirus vector can express intron-containing genes, *J. Virol., 61:* 1761–1764.

Kawanishi, C. Y., Summers, M. D., Stoltz, D. B., and Arnott, H. J. (1972). Entry of an insect virus *in vivo* by fusion of viral envelope and microvillus membrane, *J. Invertebr. Pathol., 20:* 104–108.

Keddie, B. A. and Volkman, L. E. (1985). Infectivity difference between the two phenotypes of *Autographa californica* nuclear polyhedrosis virus: Importance of the 64K envelope glycoprotein, *J. Gen. Virol., 66:* 1195–2000.

Keddie, B. A., Apointe, G. W., and Volkman, L. E. (1989). The pathway of infection of *Autographa californica* nuclear polyhedrosis virus in an insect host, *Science, 242:* 1728–1730.

Kelly, D. C. and Wang, X. (1981). The infectivity of nuclear polyhedrosis virus DNA, *Ann. Virol.* (Inst. Pasteur), *132E:* 247–259.

Kelly, D. C. (1982). Baculovirus replication, *J. Gen. Virol., 63:* 1–13.

Kelly, D. C., Brown, D. A., Ayres, M. D., Allen, C. J., and Walker, I. O. (1983). Properties of the major nucleocapsid protein of *Heliothis zea* singly enveloped nuclear polyhedrosis virus, *J. Gen. Virol., 64:* 399–408.

Kitts, P. A., Ayres, M. D., and Possee, R. D. (1990). Linearization of baculovirus DNA enhances the recovery of recombinant virus expression vectors, *Nucleic Acid Res, 18:* 5667–5672.

Kovacs, G. R., Guarino, L. A., and Summers, M. D. (1991). Novel regulatory properties of the IE1 and IE0 transactivators encoded by the baculovirus

Autographa californica multicapsid nuclear polyhedrosis virus, *J. Virol., 65:* 5281–5288.

Krappa, R. and Knebel-Morsdorf, D. (1991). Identification of the very early transcribed baculovirus gene PE-38, *J. Virol., 65:* 805–812.

Kuzio, J. and Faulkner, P. (1984). Regions of repeated DNA in the genome of *Choristoneura fumiferana* nuclear polyhedrosis virus, *Virology, 139:* 185–188.

Kuzio, J., Rohel, D. Z., Curry, C. J., Krebs, A., Carstens, E. B., and Faulkner, P. (1984). Nucleotide sequence of the p10 polypeptide gene of *Autographa californica* nuclear polyhedrosis virus, *Virology, 139:* 414–418.

Kuzio, J., Jaques, R., and Faulkner, P. (1989). Identification of p74, a gene essential for virulence of baculovirus occlusion bodies, *Virology, 173:* 759–763.

Lanford, R. E. (1988). Expression of Simian virus 40 T antigen in insect cells using a baculovirus expression vector, *Virology, 167:* 72–81.

Leisy, D. J., Rohrmann, G. F., and Beaudreau, G. S. (1984). Conservation of genome organization in two multicapsid nuclear polyhedrosis viruses, *J. Virol., 52:* 699–702.

Leisy, D. J., Rohrmann, G. F., Nesson, M., and Beaudreau, G. S. (1986). Nucleotide sequence and transcriptional mapping of the *Orgyia pseudotsugata* multicapsid nuclear polyhedrosis virus p10 gene, *Virology, 153:* 157–167.

Livingstone, C. and Jones, I. (1989). Baculovirus expression vectors with single strand capability, *Nucleic Acids Res., 17:* 2366.

Lu, A. and Carstens, E. B. (1991). Nucleotide sequence of a gene essential for viral DNA replication in the baculovirus *Autographa californica* nuclear polyhedrosis virus, *Virology, 181:* 336–347.

Lubbert, H. and Doerfler, W. (1984). Transcription of overlapping sets of RNAs from the genome of *Autographa californica* nuclear polyhedrosis virus: A novel method for mapping RNAs, *J. Virol., 52:* 255–265.

Luckow, V. A. and Summers, M. D. (1988). Trends in the development of baculovirus expression vectors, *Bio/Technology, 6:* 47–55.

Luckow, V. A. and Summers, M. D. (1989). High level expression of nonfused foreign genes with *Autographa californica* nuclear polyhedrosis virus expression vectors, *Virology, 170:* 31–39.

Luckow, V. A. (1991). Cloning and expression of heterologous genes in insect cells with baculovirus vectors, *Recombinant DNA Technology and Applications* (A. Prokop, R. K. Bajpai, and C. S. Ho, eds.), McGraw-Hill, New York, pp. 97–153.

Mackinnon, E. A., Henderson, J. F., Stoltz, D. B., and Faulkner, P. (1974). Morphogenesis of nuclear polyhedrosis virus under conditions of prolonged passage in vitro, *J. Ultrastr. Res., 49:* 419–435.

Maeda, S. (1989). Increased insecticidal effect by a recombinant baculovirus carrying a synthetic duiretic hormone gene, *Biochem. Biophys. Res. Commun., 165:* 1177–1183.

Maeda, S., Kamita, S. G., and Kataoka, H. (1991a). The basic DNA-binding protein of *Bombyx mori* nuclear polyhedrosis virus: The existence of an additional arginine repeat, *Virology, 180:* 807–810.

Maeda, S., Volrath, S. L., Hanzlik, T. N., Harper, S. A., Majima, K., Maddox, D. W., Hammock, B. D., and Fowler, E. (1991b). Insecticidal effects of an insect specific neurotoxin expressed by a recombinant baculovirus, *Virology, 184:* 777–780.

Mann, S. G. and King, L. A. (1989). Efficient transfection of insect cells with baculovirus DNA using electroporation, *J. Gen. Virol., 70:* 3501–3505.

Martens, J. W. M., Honee, G., Zuidema, D., van Lent, J. W. M., Visser, B., and Vlak, J. M. (1990). Insecticidal activity of a bacterial crystal protein expressed by a recombinant baculovirus in insect cells, *Applied and Environmental Microbiology, 56:* 2764–2770.

Matsuura, Y., Possee, R. D., Overton, H. A., and Bishop, D. H. L. (1987). Baculovirus expression vectors: the requirements for high level expression of proteins, including glycoproteins, *J. Gen. Virol., 68:* 1233–1250.

McClintock, J. T. and Dougherty, E. M. (1988). Restriction mapping of *Lymantria dispar* nuclear polyhedrosis virus DNA: Localization of the polyhedrin gene and identification of four homologous regions, *J. Gen. Virol., 69:* 2303–2312.

McCutchen, B. F., Choudary, P. V., Crenshaw, R., Maddox, D., Kamita, S. G., Palekar, N., Volrath, S., Hammock, B. D., and Maeda, S. (1991). Development of a recombinant baculovirus expressing an insect-selective neurotoxin: Potential for pest control, *Bio/Technology, 9:* 848–852.

Medin, J. A., Hunt, L., Gathy, K., Evans, R. K., and Coleman, M. S. (1990). Efficient, low-cost protein factories: Expression of human adenosine deaminase in baculovirus-infected insect larvae, *Proc. Natl. Acad. Sci. USA, 87:* 2760–2764.

Merryweather, A. T., Weyer, U., Harris, M. P. G., Hirst, M., Booth, T., and Possee, R. D. (1990). Construction of genetically engineered baculovirus insecticides containing the *Bacillus thuringiensis* subsp. kurstaki HD-76 δ endotoxin, *J. Gen. Virol., 71:* 1535–1544.

Miller, L. K., Jewell, J. E., and Browne, D. (1981). Baculovirus induction of a DNA polymerase, *J. Virol., 40:* 305–308.

Miller, L. K., Adang, M. J., and Browne, D. (1983). Protein kinase activity associated with the extracellular and occluded forms of the baculovirus *Autographa californica* nuclear polyhedrosis virus, *J. Virol., 46:* 275–278.

Miller, L. M. (1984). Exploring the gene organization of baculoviruses, *Methods in Virology, 8:* 227–258.

Miller, L. K. (1988). Baculoviruses as gene expression vectors, *Ann. Rev. Microbiol., 42:* 177–199.

Muller, R., Pearson, M. N., Russell, R. L. Q., and Rohrmann, G. F. (1990). A capsid-associated protein of the multicapsid nuclear polyhedrosis virus of *Orgyia pseudotsugata:* Genetic location, sequence, transcriptional mapping, and immunocytochemical characterization, *Virology, 176:* 133–144.

Nagata, M. and Tanada, Y. (1983). Origin of an alkaline protease associated with the capsule of a granulosis virus of the armyworm, *Pseudaletia unipuncta* (Haworth), *Arch. Virol., 76:* 245–256.

O'Reilly, D. R., Crawford, A. M., and Miller, L. K. (1989). Viral proliferating cell nuclear antigen, *Nature, 337:* 606.

O'Reilly, D. R. and Miller, L. K. (1989). A baculovirus blocks insect molting by producing ecdysteroid UDP-glucosyl transferase, *Science, 245:* 1110–1112.

O'Reilly, D. R. and Miller, L. K. (1990). Regulation of expression of a baculovirus ecdysteroid UDP-glucosyltransferase gene, *J. Virol., 64:* 1321–1328.

Oellig, C., Happ, B., Muller, T., and Doerfler, W. (1987). Overlapping sets of viral RNAs reflect the array of polypeptides in the EcoRI J and N fragments (map positions 81.2–85.0) of the *Autographa californica* nuclear polyhedrosis virus genome, *J. Virol., 61:* 3048–3057.

Ollo, R. and Maniatis, T. (1987). *Drosophila* Kruppel gene product produced in a baculovirus expression system is a nuclear phosphoprotein that binds to DNA, *Proc. Natl. Acad. Sci. USA, 84:* 5700–5704.

Ooi, B. G. and Miller, L. K. (1988). Regulation of host RNA levels during baculovirus infection, *Virology, 166:* 515–523.

Ooi, B. G., Rankin, C., and Miller, L. K. (1989). Downstream sequences augment transcription from the essential initiation site of a baculovirus polyhedrin gene, *J. Mol. Biol., 210:* 721–736.

Ooi, B. G. and Miller, L. K. (1990). Transcription of the baculovirus polyhedrin gene reduces the levels of an antisense transcript initiated downstream, *J. Virol., 64:* 3126–3129.

Ooi, B. G. and Miller, L. K. (1991). The influence of antisense RNA on transcriptional mapping of the 5′ terminus of a baculovirus RNA, *J. Gen. Virol., 72:* 527–534.

Page, M. J. (1989). p36C: An improved baculovirus expression vector for producing high levels of mature recombinant proteins, *Nucleic Acids Res., 17:* 454.

Peakman, T., Page, M., and Gewert, D. (1989). Increased recombinational efficiency in insect cells irradiated with short wavelength ultraviolet light, *Nucl. Acids Res., 17:* 5403.

Pearson, M. N., Russell, R. Q., Rohrmann, G. F., and Beaudreau, G. S. (1988). p39, a major baculovirus structural protein: Immunocytochemical characterization and genetic location, *Virology, 167:* 407–413.

Pennock, G. D., Shoemaker, C., and Miller, L. K. (1984). Strong and regulated expression of *Escherichia coli* β-galactosidase in insect cells with a baculovirus vector, *Mol. Cell. Biol., 4:* 399–406.

Possee, R. D., Weyer, U., and King, L. A. (1990). Recombinant antigen production using baculovirus expression vectors, *Control of Viral Diseases* (N. J. Dimmock, P. D. Griffiths, and C. R. Madley, eds.), Cambridge University Press, Cambridge, pp. 53–76.

Possee, R. D., Bonning, B. C., and Merryweather, A. T. (1991). Expression of proteins with insecticidal activities using baculovirus vectors, *Ann. N.Y. Acad. Sci., 646:* 234–239.

Potter, K. N., Faulkner, P., and MacKinnon, E. A. (1976). Strain selection during serial passage of *Trichoplusia ni* nuclear polyhedrosis virus, *J. Virol., 18:* 1040–1050.

Price, P. M., Reichelderfer, C. F., Johansson, B. E., and Kilbourne, E. D. (1989). Complementation of recombinant baculoviruses by coinfection with wild-type virus facilitates production in insect larvae of antigenic proteins of hepatitis B virus and influenza virus, *Proc. Natl. Acad. Sci. USA, 86:* 1453–1456.

Quant-Russell, R. L., Pearson, M. N., Rohrmann, G. F., and Beaudreau, G. S. (1987). Characterization of baculovirus p10 synthesis using monoclonal antibodies, *Virology, 160:* 9–19.

Rankin, C., Ladin, B. F., and Weaver, R. F. (1986). Physical mapping of temporally regulated, overlapping transcripts in the region of the 10K protein gene in *Autographa californica* nuclear polyhedrosis virus, *J. Virol., 57:* 18–27.

Roberts, T. E. and Faulkner, R. (1989). Fatty acid acylation of the p67 glycoprotein of a baculovirus: *Autographa californica* nuclear polyhedrosis virus, *Virology, 172:* 377–381.

Rohrmann, G. F. (1986a). Polyhedrin structure, *J. Gen. Virol., 67:* 1499–1513.

Rohrmann, G. F. (1986b). Evolution of occluded baculoviruses, *The Biology of Baculoviruses* (R. R. Granados and B. A. Federici, eds.), CRC Press, Boca Raton, Florida, pp. 203–215.

Rohrmann, G. (1991). Baculovirus structural proteins, *J. Gen. Virol., 73:* 749–761.

Rubenstein, R. and Polson, A. (1983). Midgut and viral associated proteases of *Heliothis armigera, Intervirol., 19:* 16–25.

Russell, R. L. Q. and Rohrmann, G. F. (1990). A baculovirus polyhedron envelope protein: Immunogold localization in infected cells and mature polyhedra, *Virology, 174:* 177–184.

Russell, R. L. Q., Pearson, M. N., and Rohrmann, G. F. (1991). Immunoelectron microscopic examination of *Orgyia pseudotsugata* multicapsid nuclear polyhedrosis virus-infected *Lymantria dispar* cells: Time course and localization of major polyhedron-associated proteins, *J. Gen. Virol., 72:* 275–283.

Russell, R. L. Q. and Rohrmann, G. F. (1991). The p6.5 gene region of a nuclear polyhedrosis virus in *Orgyia pseudotsugata:* DNA sequence and transcription analysis of four late genes, *J. Gen. Virol., 71:* 551–560.

Sarvari, M., Csikos, G., Sass, M., Gal, P., Schumaker, V. N., and Zavodszky, P. (1990). Ecdysteroids increase the yield of recombinant protein produced in baculovirus insect cell expression system, *Biochem. Biophys. Res. Commun., 167:* 1154–1161.

Smith, G. E., Fraser, M. J., and Summers, M. D. (1983). Molecular engineering of the *Autographa californica* nuclear polyhedrosis virus genome—deletion mutations within the polyhedrin gene, *J. Virol., 46:* 584–593.

Stewart, L. M. D., Hirst, M., Ferber, M. L., Merryweather, A. T., Cayley, P. J., and Possee, R. D. (1991). Construction of an improved baculovirus insecticide containing an insect-specific toxin gene, *Nature, 352:* 85–88.

Stoltz, D. B., Pavan, C., and Dacunha, A. B. (1973). Nuclear polyhedrosis virus: A possible example of *de novo* intranuclear membrane morphogenesis, *J. Gen. Virol., 19:* 145–150.

Summers, M. D., Volkman, L. E., and Hseih, C. H. (1978). Immunoperoxidase detection of baculovirus antigens in insect cells, *J. Gen. Virol.,* *40:* 545–557.

Summers, M. D. and Smith, G. E. (1987). *A Manual of Methods for Baculovirus Vectors and Insect Cell Culture Procedures,* Texas Agriculture Experiment Station, Bull. 1555, College Station, Texas.

Summers, M. D. (1989). Recombinant proteins expressed by baculovirus vectors, *Concepts in Viral Pathogenesis III* (A. L. Notkins and M. B. A. Oldstone, eds.), Springer-Verlag, New York, pp. 77–86.

Thielmann, D. A. and Stewart, S. (1991). Identification and characterization of the IE-1 gene of *Orgyia pseudotsugata* multicapsid nuclear polyhedrosis virus, *Virology, 180:* 492–508.

Thiem, S. M. and Miller, L. K. (1989a). Identification, sequence and transcriptional mapping of the major capsid protein gene of the baculovirus *Autographa californica* nuclear polyhedrosis virus, *J. Virol., 63:* 2008–2018.

Thiem, S. M. and Miller, L. K. (1989b). A baculovirus gene with a novel transcription pattern encodes a polypeptide with a zinc finger and a leucine zipper, *J. Virol., 63:* 4489–4497.

Thiem, S. M. and Miller, L. K. (1990). Differential gene expression mediated by late, very late and hybrid baculovirus promoters, *Gene, 91:* 87–94.

Tomalski, M. D., Wu, J., and Miller, L. K. (1988). The location, sequence, transcription, and regulation of a baculovirus DNA polymerase gene, *Virology, 167:* 591–600.

Tomalski, M. D., Eldridge, R., and Miller, L. K. (1991). A baculovirus homolog of a Cu/Zn superoxide dismutase gene, *Virology, 184:* 149–161.

Tomalski, M. D. and Miller, L. K. (1991). Insect paralysis by baculovirus-mediated expression of a mite neurotoxin gene, *Nature, 352:* 82–85.

Tweeten, K. A., Bulla, L. A. Jr., and Consigli, R. A. (1980). Characterization of an extremely basic protein derived from granulosis virus nucleocapsids, *J. Virol., 33:* 866–876.

van der Wilk, F., van Lent, J. W., and Vlak, J. M. (1987). Immunogold detection of polyhedrin, p10 and virion antigens in *Autographa californica* nuclear polyhedrosis virus infected *Spodoptera frugiperda* cells, *J. Gen. Virol., 68:* 2615–2623.

van Lent, J. W. M., Groenen, J. T. M., Kling-Roode, E. C., Rohrmann, G. F., Zuidema, D., and Vlak, J. M. (1990). Localization of the 34kDa polyhedron envelope protein in *Spodoptera frugiperda* cells infected with *Autographa californica* nuclear polyhedrosis virus, *Archives of Virology, 111:* 103–114.

Vialard, J., Lalumiere, M., Vernet, T., Briedis, D., Alkhatib, G., Henning, D., Levin, D., and Richardson, C. (1990a). Synthesis of the membrane fusion and hemagglutinin proteins of measles virus, using a novel baculovirus vector containing the β-galactosidase gene, *J. Virol., 64:* 37–50.

Vialard, J. E., Yuen, L., and Richardson, C. D. (1990b). Identification and characterization of a baculovirus occlusion body glycoprotein which resembles spheroidin, an entomopoxvirus protein, *J. Virol., 64:* 5804–5811.

Vlak, J. M. and Smith, G. E. (1982). Orientation of the genome of *Autographa californica* nuclear polyhedrosis virus: A proposal, *J. Virol., 41:* 1118–1121.

Vlak, J. M., Klinkenberg, F. A., Zaal, K. J. M., Usmay, M., Klinge-Roode, E. C., Geervliet, J. B. F., Roosien, J., and van Lent, J. W. M. (1988). Functional studies on the p10 gene of *Autographa californica* nuclear polyhedrosis virus using a recombinant expressing a p10-β-galactosidase fusion gene, *J. Gen. Virol., 69:* 765–776.

Vlak, J. M., Schouten, A., Usmany, M., Belsham, G. J., Klinge-Roode, E. C., Maule, A. J., van Lent, J. W. M., and Zuidema, D. (1990). Expression of cauliflower mosaic virus gene I using a baculovirus vector based upon the p10 gene and a novel selection method, *Virology, 179:* 312–320.

Volkman, L. E., Goldsmith, P. A., Hess, R. T., and Faulkner, P. (1984). Neutralization of budded *Autographa californica* NPV by a monoclonal antibody: Identification of the target antigen, *Virology, 133:* 354–362.

Volkman, L.E. and Goldsmith, P. A. (1985). Mechanism of neutralization of budded *Autographa californica* nuclear polyhedrosis virus by a monoclonal antibody: Inhibition of entry by adsorptive and endocytosis, *Virology, 143:* 185–195.

Volkman, L. E. (1986). The 64K envelope protein of budded *Autographa californica* nuclear polyhedrosis virus, *Curr. Top. Microbiol. Immunol., 131:* 103–118.

Volkman, L. E. and Knudson, D. L. (1986). In vitro replication of baculoviruses, *The Biology of Baculoviruses* (R. R. Granados and B. A. Federici, eds.), CRC Press, Boca Raton, Florida, pp. 109–127.

Volkman, L. E. and Keddie, B. A. (1990). Nuclear polyhedrosis virus pathogenesis. *Seminars in Virology,* UC Berkeley, 1: 249–256.

Volkman, L. E. and Zaal, K. J. M. (1990). *Autographa californica M* nuclear polyhedrosis virus: Microtubules and replication, *Virology, 175:* 292–302.

Wang, X., Ooi, B. G., and Miller, L. K. (1991). Baculovirus vectors for multiple gene expression and for occluded virus production, *Gene, 100:* 131–137.

Weyer, U., Knight, S., and Possee, R. D. (1990). Analysis of very late gene expression by *Autographa californica* nuclear polyhedrosis virus and the further development of multiple gene expression, *J. Gen. Virol., 71:* 1525–1534.

Whitford, M., Stewart, S., Kuzio, J., and Faulkner, P. (1989). Identification and sequence analysis of a gene encoding gp67, an abundant envelope glycoprotein of the baculovirus *Autographa californica* nuclear polyhedrosis virus, *J. Virol., 63:* 1393–1399.

Whitt, M. A. and Manning, J. E. (1988). A phosphorylated 34 kDa protein and a subpopulation of polyhedrin are thiol linked to the carbohydrate layer surrounding a baculovirus occlusion body, *Virology, 163:* 33–42.

Williams, G. V., Rohel, D. Z., Kuzio, J., and Faulkner, P. (1989). A cytopathological investigation of *Autographa californica* nuclear polyhedrosis virus (AcNPV) p10 gene function using insertion/deletion mutants, *J. Gen. Virol., 70:* 187–202.

Wilson, M. E. and Consigli, R. A. (1985a). Functions of a protein kinase activity associated with purified capsids of the granulosis virus infecting *Plodia interpunctella, Virology, 143:* 526–535.

Wilson, M. E. and Consigli, R. A. (1985b). Characterization of a protein kinase associated with purified capsids of the granulosis virus infecting *Plodia interpunctella, Virology, 143:* 516–525.

Wilson, M. and Miller, L. K. (1986). Changes in the nucleoprotein complexes of a baculovirus DNA during infection, *Virology, 151:* 315–328.

Wilson, M. E., Mainprize, T. H., Freisen, P. D., and Miller, L. K. (1987). Location, transcription and sequence of a baculovirus gene encoding a small arginine rich polypeptide, *J. Virol., 61:* 661–666.

Wilson, M. E. and Price, K. H. (1988). Association of *Autographa californica* nuclear polyhedrosis virus with the nuclear matrix, *Virology, 167:* 233–241.

Wu, J. and Miller, L. K. (1989). Sequence, transcription and translation of a late gene of the *Autographa californica* nuclear polyhedrosis virus encoding a 38.4K polypeptide, *J. Gen. Virol., 70:* 2449–2459.

Yuen, L., Dionne, J., Arif, B., and Richardson, C. (1990). Identification and sequencing of the spheroidin gene of *Choristoneura biennis* entomopoxvirus, *Virology, 175:* 427–433.

Zuidema, D., Klinge-Roode, E. C., van Lent, J. W. M., and Vlak, J. M. (1989). Construction and analysis of an *Autographa californica* nuclear polyhedrosis virus mutant lacking the polyhedron envelope, *Virology, 173:* 98–108.

Zuidema, D., Schouten, A., Usmany, M., Maule, A. J., Belsham, G., Roosien, J., Klinge-Roode, E. C., van Lent, J. W. M., and Vlak, J. M. (1990). Expression of cauliflower mosaic virus gene 1 in insect cells using a novel polyhedrin based baculovirus expression vector, *J. Gen. Virol., 71:* 2201–2210.

Zu Putlitz, J., Kubasek, W. L., Duchene, M., Marget, M., von Specht, B., and Domdey, H. (1990). Antibody production in baculovirus infected insect cells, *Bio/Technology, 8:* 651–654.

3

Bioreactor Design for Insect Cell Cultivation: A Review

Janusz J. Malinowski   *Polish Academy of Sciences, Gliwice, Poland*

Andrew J. Daugulis   *Queen's University, Kingston, Ontario, Canada*

3.1 INTRODUCTION

Many commercially important products of the cells of higher organisms may be produced *in vitro* by cell culture techniques. The products that could be produced by the culture of the three classes of cells of higher organisms are: (i) therapeutics such as vaccines, hormones, functional proteins, and peptides in vertebrate cells; (ii) viruses such as nuclear polyhedrosis virus (NPV), as well as therapeutic proteins in invertebrate cells, (iii) alkaloids in plant cells. An assessment of the predictions of the 1980s has led to the conclusion that the capability of genetically engineered bacteria and lower eukaryotes to produce most commercially valuable proteins was largely overestimated. It now appears that a significant proportion of economically attractive animal proteins must be produced on a large scale using higher eukaryotic cells. This is providing a strong impetus for research and development activities focusing on more efficient cultivation of animal cells *in vitro*.

During the last few years, significant developments in improved and scaled-up systems of insect cell culture have emerged. Earlier attempts at large-scale insect cell tissue culture were motivated by the possibility of mass-producing viral insecticides. Insect viruses are attractive as insecticides because they are very specific for a limited range of insects and nonpathogenic to vertebrates. Insects cause considerable damage in agricultural production worldwide, and one environmentally sound alternative to chemical pest control is the use of insect pathogens, including viruses (Burges, 1981; Martignoni, 1984; Kurstak, 1982; Podgwaite, 1985). Among the viruses that are pathogenic to insects are baculoviruses. These viruses are found almost exclusively among insects and they are a threat to neither human health nor to environment.

The current interest in insect cell culture, however, is due to the recent development of baculovirus vectors for the expression of many foreign genes at high levels (Luckow and Summers, 1988). One of the major advantages of this invertebrate virus expression system over bacterial, yeast and mammalian expression systems is the very abundant expression of recombinant proteins, which are in many cases similar to their authentic counterparts. This is due, in large part, to the ability of the insect cell hosts to properly posttranslationally modify the recombinant protein products. Among the most visible recent applications of this system is the development of the first candidate vaccine for AIDS (van Brunt, 1987), following the expression of the recombinant HIV envelope proteins in insect cell culture (Madisen et al., 1987; Hu, Kosowski and Schaaf, 1987).

Insect cell systems of interest for large-scale applications are not limited only to the cell types that can serve as hosts to baculoviruses (primarily cells of lepidopteran), but include also cells of dipteran insects (mosquitoes) that transmit many important diseases or are the materials for the genetic and molecular biological studies of the fruit fly, *Drosophila*. It is clear that efficient methods are needed in the commercial exploitation of insect cell cultures.

Viral insecticides are a present day example of products manufactured *in vitro* from insect cells using whole insect larvae (Shapiro, 1986; Shieh and Bohmfalk, 1980). However, there are many drawbacks in the *in vitro* process: it requires special pathogen-free, insect-growing facilities as well as high labor and operating costs (Khachatourians, 1986). Additionally, the product is impure, containing many different contaminants (e.g., insect cuticle, proteins) which can be highly allergenic to humans and must be removed from the virus before incorporation into the final product. The process itself is not easily compatible with automatic control and is difficult to scale up. Thus, the *in vitro* process of insect cell culture for the large scale production of insect pathogenic viruses and recombinant proteins appears to be more suitable.

Within the framework of insect cell culture techniques, the growth of insect cells in monolayers or stationary culture may be unsuitable for large-scale production due to the decrease in the surface to volume ratio with an increase in the scale of operation. This may make capital costs for large systems of this type prohibitive (Pollard and Khosrovi, 1978). Therefore, the most promising process development approaches are being oriented towards cultivation of insect cells in suspension, using bioreactor technology currently applied to microbial and mammalian cell cultures.

The purpose of this review is to summarize aspects of bioreactor technology that have been or could be applied to insect cell cultures with an emphasis on bioreactor design considerations.

3.2 LIMITATIONS IN BIOREACTOR DESIGN FOR INSECT CELL CULTURES

By the late 1950s, insect cell culture was developing into a useful tool for virus study. Rapid developments in the field of insect cell culture, in the period between 1967 and 1974, are evident by the fact that of the approximately 90 cell lines used today, about 80 were described in this time interval. However, the development of culture techniques has often been undertaken by those intending to study viruses, and little attention has been given to the technology associated with the cultivation of insect cells. Growing interest in the field use of viral insecticides in pest control has stimulated investigations into effective methods of larger scale insect cell cultivation.

Cells originally grew in stationary cell culture flasks. Vaughn (1968) reported in 1967 the first successful attempt at growing insect cells (*Antheraea* cell line) in suspension. Lengyel, Spradling and Penman (1975), working with fruit fly cells, (*Drosophila melanogaster*) adapted the cells to grow in roller bottles and then in spinner flasks. The maximum densities were 1 to 4×10^6 cells/ml, but densities as high as 10^7 were reached. This same group of researchers also adapted mosquito (*Aedes albopictus*) cells to grow in suspension.

However, there are obvious advantages to using bioreactors of standard design for insect cell growth with only minor modification. Hink's group pioneered attempts in this direction since the mid 1970s, working with cabbage looper cells (*Trichoplusia ni*, TN-368) as a useful system for producing *Autographa californica* nuclear polyhedrosis virus (AcNPV). The first step was to place the TN-368 cells in water jacketed spinner flasks with a working volume of 100 ml (Hink and Strauss, 1976). The cells grew in spinner flasks but they did not reach as high a density as in stationary culture. A cell density of 8×10^5 to 1×10^6 cells/ml was obtained with initial populations of $1–2 \times 10^5$ cells/ml. In an effort to improve on this result, Hink and Strauss determined that O_2 tension, cell clumping and medium pH are important growth regulating factors. They found that cell growth was increased by aeration (i.e., sparging air through the medium rather than passing a stream of air above the medium surface). The best cell growth was achieved when filtered air was passed at a rate of 5 cm^3/min into the flask through a stainless steel tube extending to the bottom of the culture vessel. The pH was adjusted and methylcellulose (0.1%) was added to the medium to prevent cell clumping. These conditions resulted in a maximum density of 3×10^6 cells/ml in 72 hrs.

Another system tested was a 400 ml Vibromixer™ glass bioreactor (Hink and Strauss, 1976). The Vibromixer, introduced by Chemap, is composed of a horizontal disk with conical apertures attached to a vertical shaft that rapidly oscillates up and down. The Vibromixer circulates the medium in a vertical direction providing mixing at fairly low shear. The bioreactor was originally designed for mammalian cell cultures which prefer the axial type of flow pattern (Prokop and Rosenberg, 1989). However, in the case of insect cell cultivation, this system gave even slower growth, and cells appeared to have unhealthy morphology. These phenomena were attributed to mixing-associated shear in the bioreactors.

Various problems arose from the first attempts of Hink's group (Hink and Strauss, 1980; Hink, 1982) to increase the scale of operation. The New Brunswick, model MF-205 4 liter jar bioreactor equipped with marine impellers was used. Air or O_2 was sparged from the bottom up through the medium at a working volume of 2 to 3 liter. The increase in volume necessitated agitating and aerating the cultures at faster rates, which caused foaming of the medium. The impeller speed was increased from 100 rpm (used for spinner flask) to 200 rpm, and the aeration rate from 5 ml/min (used for the spinners) to 750 ml/min (after 48 hr at 125 ml/min). With addition of 0.02% antifoam, the cell density reached 1.75 $\times 10^6$ cells/ml within 72 hr; without antifoam, it took 96 hr to grow to the same density. The beneficial effect of silicon-based antifoam on cell growth was possibly due to reduced cell loss resulting from entrainment in foam. To prevent cell damage at higher impeller speed, the medium viscosity was enhanced by increasing the methylcellulose content from 0.1%, previously used in spinner flasks, to 0.3%.

The same researchers (Hink and Strauss, 1980; Hink, 1982) also found that the level of dissolved oxygen (DO) affected the growth rate of insect cells in the stirred bioreactors. When DO was allowed to drop freely from an initial value of 100% to 2% within 24 hr, the cells stopped growing. In contrast, the aeration necessary to maintain the DO at 100% caused vacuolation of cells and precipitate formation after the cell growth had levelled off. Cultures maintained at 50% DO by air sparging grew just as well as those kept at 100% DO by oxygen sparging. Under these conditions, it was possible to attain final cell densities of 5×10^6 cells/ml in 5 days, with initial densities of 2×10^6 cells/ml, and a corresponding doubling time of only 14 hr during exponential growth. Therefore, it seems that a combination of effective aeration, viscous inert additives, and agitation with low-shear impellers (e.g., propeller type) may enhance the growth of insect cell culture in bioreactors of standard stirred tank design.

The main technological constraint to large volume cultivation of insect cells in suspension is the supply of sufficient oxygen without aggressive sparging and stirring. In comparison to microbial cells, insect cells are very sensitive to shear force due to their relatively large size (10–20 μm range) and lack of a cell wall. A semiquantitative indication of the shear sensitivity of insect cells is given by Hink (1982) who found that growth of *T.ni* cells in a 3 liter fermentor stopped at a stirrer speed of 220 rpm in medium containing 0.1% methylcellulose, but continued in 0.3% methylcellulose at this agitation rate. From these experiments, a critical shear stress of the order of 1.5 Nm^{-2} was deduced (Tramper et al., 1986). Tramper and co-workers (Tramper, et al., 1986) assessed systematically the shear sensitivity of *Spodoptera frugiperda* cells cultivated in a 1 liter round bottomed fermentor (Applikon) equipped with a marine impeller, in medium containing 0.1% methylcellulose. In two different experiments, the viability of cells started to decrease at 220 rpm and 510 rpm. These stirrer speeds correspond roughly to shear stresses of 1 Nm^{-2} and 3 Nm^{-2} respectively. A more accurate investigation of shear stress effects on cell viability carried out using a Haake rotaviscometer confirmed the critical shear stress at which the number of viable cells declines to be between 1 and 4 Nm^{-2}. Maiorella et al. (1991) found that the sensitivity of animal cells to hydrodynamic stress is increased in serum-free and low protein media. They studied the effect of laminar shear on cell viability in crossflow microfiltration systems at several flow rates. The loss of viability by Sf-9 insect cells was observed above an average shear rate of 3000 s^{-1}. This critical shear rate corresponds to an average shear stress on the cells of 2.5 Nm^{-2}. In all experiments, the protective agent (Pluronic F-68) was added to the culture media.

Recently, Goldblum et al. (1990) reported on the extreme sensitivity of *T.ni* (TN-368) and *S. frugiperda* (Sf-9) cell lines in laminar shear stress. The suspended cells were subjected to well-defined shear stress in a modified Weissenberg R-16 rheogoniometer. Significant cell lysis at shear stresses of 0.1 Nm^{-2} with TN-368 cells and 0.59 Nm^{-2} with Sf-9 cells after only 5 min in medium without any additives was observed. This is one order of magnitude below the value determined by Tramper et al. (1986) and Maiorella et al. (1991) for Sf-9 cells in media with 0.1% methylcellulose or 0.1% Pluronic F-68 added. This shows some protective effects of high viscosity agents on the cells' resistance to shear stress.

Because cell physiology can be modulated by the environment, it is expected that fluid shear stresses will influence the efficiency of the insect cell as a host for protein production from recombinant DNA. There is evidence obtained from mammalian cell culture that shear stresses could modulate protein synthesis (Shuler et al., 1990).

Oxygen requirements of insect cells are high (15 to 45 μg O_2/mg cell-hr) (Stockdale and Gardiner, 1976) compared to mammalian cells (1.7 to 19 μg O_2/mg cell-hr) (Streett and Hink, 1978; Glacken, Fleischaker and Sinskey, 1983). Streett and Hink (1978) found that these requirements nearly double for *T.ni* cells after being infected with AcNPV baculoviruses. Also, differentiated cells, which might be desirable host cells, have elevated oxygen demand (Shuler et al., 1990). In order to satisfy the oxygen demand to systems employing insect cells, the standard approaches of microbial fermentation technology such as agitation and aeration by sparging may be too aggressive.

Most recently, Zhang, Handa-Corrigan and Spier (1992) have extensively studied the problem associated with oxygen transfer properties of bubbles in animal cell culture media. From an oxygen transfer point of view, micron-sized bubbles ($\sim 100 \mu m$ diameter) in media containing Pluronic F-68 polyol are recommended while keeping serum concentration to a minimum. Employing an appropriate sparger, bubbles of the desirable size and shape can be generated into culture media.

The effect of airflow on viability of *S. frugiperda* was studied by Tramper et al. (1986) using a bubble column reactor (height 0.18 m, inside diameter 0.035 m). The experiment was performed at different gas flows and using different air spargers. They found that the first order death rate constant for these insect cells was proportional to the air flow rate. The effect of bubble size was much less pronounced. The loss of viability due to direct air sparging was explained in terms of an adherence of cells to the bubble/liquid interface and bursting of bubbles at the surface of the suspension. The estimated shear stress such an adhered cell could experience (625 Nm^{-2}) is roughly two orders of magnitude above the critical value of 1 Nm^{-2}. These findings may explain unsuccessful efforts to grow the cells in an airlift type bioreactor (Tramper et al., 1986). Alternative oxygenation strategies, such as membrane bioreactors or diffusion of gas through semipermeable tubing, were recommended (Tramper et al., 1986; Tramper and Vlak, 1986).

Direct sparging of gas has also been shown to be detrimental to the viability of mammalian cells (Tramper and Vlak, 1986), and the damaging effect of air bubble flow has appeared to be limited only to the region of bubble disengagement at the medium surface. Handa-Corrigan, Emery and Spier (1989) found that bubble column bioreactors with large height to diameter ratios ensure better cell growth due to reduced time spent by the cells in the bursting bubble zone. The importance of bubble disengagement for the loss of insect cell viability in a bubble column reactor was demonstrated by Wudtke and Schügerl (1987). Higher cell viability was observed when the free suspension surface was covered with paraffin oil to prevent bubble bursting.

Recent work (Murhammer and Goochee, 1990b; Kunas and Papoutsakis, 1990) indicates that gas sparging is the major shear related problem for insect and mammalian cell cultures. No adverse effects on the growth of Sf-9 cells were observed in a 3 liter bioreactor (Applicon, Foster City, CA) agitated up to 300 rpm with a flat six blade impeller, or 500 rpm with a scoping marine impeller. The bioreactor was oxygenated by surface aeration. Kunas and Papoutsakis (1990) have shown that two fluid mechanical mechanisms can cause hybridoma cell damage and growth retardation in agitated bioreactors. The first is associated with vortex formation, bubble entrainment, and breakup. In the absence of a vortex and bubble entrainment, cells can be damaged only with extremely high agitation rates (e.g., 700 rpm in a 1–2 liter bioreactor). It can be expected that similar mechanisms are responsible for insect cell damage in agitated bioreactors. Kunas and Papoutsakis (1990) have concluded that with proper reactor design, direct sparging at high agitation rates may be feasible without detrimental effects on cells. What design would make this feasible is not yet clear.

Communications on unusually high shear sensitivity of insect cells (Hink, 1982; Tramper et al., 1986), on cell damage by direct air sparging (Tramper et al., 1986), and on high oxygen demand of insect cell cultures (Stockdale and Gardiner 1976; Weiss et al., 1982) make evident the need for special bioreactor designs and media additives for protecting freely suspended cells against shear related damage. These protective additives for animal cell cultures media include various derivatized celluloses and starches, pluronic polyols, protein mixtures, dextrans, polyethylene glycol (PEG) and polyvinyl alcohol (PVA) (Papoutsakis,

1991). Of these, only Pluronic F-68 has been well established as a shear protectant for use in sparged agitated biorectors and airlift reactors (Murhammer and Goochee, 1988, 1990a,b; Maiorella et al., 1988). Murhammer and Goochee (1990a) found that in insect cell (Sf-9) cultures, PEG protected cells to some extent in agitated and sparged reactors, but not at all in airlift biorectors. However, there are also emerging reports indicating some disagreement on the high shear sensitivity and the values of oxygen requirements of insect cells (Wudtke and Schügerl, 1987; Maiorella et al., 1988). Therefore, practical information on the behavior of insect cells cultivated in different culture configurations (suspension, attached growth) and in different environmental conditions (aeration, agitation) are still needed. This knowledge, together with an understanding of the biological aspects of insect cell cultivation, may create the basis for the design and implementation of large-scale cell culture.

3.3 LARGE-SCALE INSECT CELL SUSPENSION CULTIVATION BIOREACTOR CONSIDERATIONS

The growing prospects of employing insect cell cultures for the production of a variety of bioproducts, ranging from agricultural viral pesticides to human health care recombinant proteins, have prompted considerable efforts at increasing the scale of operation. The scale does not necessarily mean bioreactor size but may relate to increased product capacity, which can be enhanced either by increasing cell density, reactor volume, or both (Agathos, Jeong and Venkat, 1990). In fact, a trend toward increasing cell concentrations one or two orders of magnitude above what is achieved in conventional bioreactors for mammalian cell cultures is a promising development (Tyo and Spier, 1987). In the design of a bioreactor for insect cell cultivation, the unique characteristics of insect cells with their shear fragility, especially, must be taken into account. Over the last decade, different reactor configurations for larger scale insect cell cultures have been used.

Miltenburger and David (1980) succeeded in culturing a lepidopteran cell line (IZD-MB0503) isolated from *Mamestra brassicae* (cabbage moth) in a stirred tank bioreactor. The fermentor used was a 12 liter Biostat-S (B. Brann AG, Melsungen, Germany) equipped with a stirring propeller. The working volume of the culture was 5–10 liters. To overcome the difficulty of oxygen supply without the detrimental effect of direct air sparging, a semipermeable silicon rubber tube (outer and inner diameter 6 and 5 mm respectively) was used for aeration by diffusion. Fifteen meters of tubing was coiled around the fermentor's heating system to meet the oxygen demand of cultured cells. They obtained a final cell density of 3×10^6 cells/ml within 3–4 days, with inoculum concentration of $1–2 \times 10^5$ cells/ml. The doubling times ranged from 17 to 28 hr. An additional advantage of this oxygenation system was that foaming was avoided, which mitigated other potential problems connected with foam formation, such as cell entrainment and flotation.

This indirect aeration system was tested further by Eberhard and Schügerl (1987) during cultivation of *S. frugiperda* (IPLB-Sf) cells in a 14 liter Biostat-E fermentor fitted with a blade impeller. They confirmed that the oxygen transfer rate of the aeration system depended on the operation parameters (e.g., inlet air/oxygen pressure and agitation rate). An increase of inlet air pressure from 1.0 to 1.5 bar resulted in a 50% increase in the cell yield, and in faster growth rates. At a low agitation rate (325 rpm), the low oxygen transfer and inhomogeneity of cell suspension (possibly causing mass transfer limitation) resulted in poor cell growth (final cell density of only 1×10^6 cells/ml). By increasing the stirring speed to 50, 70, or 100 rpm, the oxygen transfer rate increased. Instead of the expected enhancement of cell yield, all fermentations with higher agitation rates (higher shear forces) gave nearly the same final cell concentration of 2.4×10^6 cells/ml. However, by employing intermediate agitation (50 rpm), and upon increasing the oxygen transfer rate by either introducing pure oxygen in the silicon tube, or using air both through the tubing and over the surface of the culture liquid, the final cell density was $4–5 \times 10^6$ cells/ml. These values were comparable with those routinely obtained in 50 ml spinner flasks, where headspace aeration was sufficient for oxygen supply. This tubing-based aeration system was recommended for the large fermenter (Eberhard and Schügerl, 1987).

Fleischaker and Sinskey (1981) also demonstrated the utility of this aeration strategy to provide oxygen to a 7.5 liter mammalian cell culture. However, this type of aeration would not be a practical solution for large vessels (Glacken, Fleischaker and Sinskey, 1983; Prokop and Rosenberg, 1989; Shuler, et al., 1990). Glacken, Fleischaker and Sinskey (1983) calculated the length of tubing that must be used to oxygenate 1000 liters of a culture of HeLa cells. The value of 100 feet of 1 inch silicon tubing obtained for mammalian cells showed that this solution may be expensive and quite unwieldly, particularly in the case of insect cell cultivation when the oxygen demand may be even higher. Shuler et al. (1990) doubted the utility of immersed silicon rubber tubing for aeration of large-scale culture reactors due to the limited length of tubing that can be used before problems with the plug flow of gas and its decreasing concentration of oxygen become apparent.

Another indirect aeration system for animal cell cultivation was proposed by Chemap Inc. (Karrer, 1988). In the Chemcell system, the bioreactor is divided into a bubble-free cultivation zone and a cell-free aeration zone. The zones are separated by a vibrating mesh, which improves the diffusion of oxygen into the system. The same mesh also works as a cell retention device in high cell density cultures.

Sulzer (ProBioTech, 1990) launched a newly designed perfusion fermenter (Spinferm). The MBR Reactor Department of Sulzer (UK) developed this design to overcome the problems arising from the extreme fragility of cells and their susceptibility to mechanical stress. Its novel feature is a cylindrical, stainless steel, mesh perfusion filter which is rotated in the vessel, and is used to keep cells in the annular space while allowing medium to be exchanged across the central cylinder. Options include a low-shear stirrer in combination with a draught tube to ensure mixing and circulation of the cells and a bubble-free aerator. This fermenter is available with working volume up to 1000 liters.

Agathos (1988) and Agathos, Jeong and Venkat (1990) studied the growth kinetics of C7-10 cells derived from the *Aedes albopictus* (forest day mosquito), for the purpose of arbovirus (antigen diagnostics) production. A 4 liter MMF-05 (New Brunswick Sci.) jar fermentor was employed, with a standard marine impeller placed close to the bottom of the vessel, and a turbine impeller placed slightly below the liquid surface. A working volume of 1–2 liters and an agitation rate of 60 rpm were used. A 5% CO_2 in air mixture in the head space of the bioreactor (surface aeration) fulfilled the oxygen demand of the culture. To prevent shear damage, 0.1% methylcellulose was added to the medium. The maximum cell concentration of 3.6×10^6 cells/ml was achieved in 200 hr.

They reported the highest cell density in spinner flask culture to be 5×10^6 cells/ml. Cell clumping, which started at a cell density of $2.5–3.0 \times 10^6$ cells/ml, was observed in this culture. This clumping phenomena may have adverse effects on cell growth. Agathos, Jeong and Venkat (1990) noted a decrease in protein systhesis by cells with an increase in cell concentration, which may reflect the negative influence of cell clumping. The cell aggregation may cause an internal mass transfer limitation that results in decreased cell viability and decreased external mass transfer area. Additionally, cell clumping is undesirable due to a form of contact inhibition that insect cells appear to exhibit (Shuler et al., 1990). This induces a reduction of host and viral DNA replication in high density cultures. The problem of cell aggregation may possibly be avoided by the proper choice of agitation speed while still taking into account the shear sensitivity of the cells. Medium additives, such as Pluronic F-68, can also inhibit cell aggregation (Papoutsakis, 1991).

Weiss et al. (1988), using serum-free medium, have successfully grown Sf-9 cells in an 8 liter modified spinner bioreactor. The air was sparged at the rate of 25 to 75 ml/min through a gas sparging ring. The impeller speed was maintained at 75 rpm. The cell density obtained ranged from 2.9 to 3.9 $\times$ 10^6 total cells/ml. This is an encouraging result on the prospects of using agitated bioreactors with a direct air sparging system for insect cell cultivation.

More recently, Tramper and co-workers (Tramper et al., 1988; Tramper et al., 1990) extended the findings of their bubble column experiments (Tramper et al., 1986) to a bubble column design approach for growth of fragile insect cells. In developing a model for the system, it was assumed that cell death, as a result of air sparging, is a first-order process, and a killing volume in which all viable cells are killed, was associated with each rising air bubble. The killing-volume hypothesis yielded useful design equations relating cell growth, cell death by aeration, and oxygen supply, in a bubble column (Tramper et al., 1988). The model was experimentally validated on a lab scale.

Another type of fermenter, offering several advantages for large scale bioprocesses with shear sensitive cultures, is an airlift bioreactor (Merchuk, 1990). Airlift reactors can be considered a type of bubble column, since these are also pneumatically agitated, but the main difference lies in the fluid flow. The bubble column is a simple vessel where a random mixing of the culture medium is caused by the rising swarm of bubbles. In contrast, in the airlift fermentor, the pattern of fluid circulation is determined by the bioreactor design. Gas is sparged at the base of the vessel, rises within a cylindrical draft tube, (the riser), and releases at the free surface of the medium. The difference in gas holdup between the riser and the downcomer region, outside the riser, is the driving force for a gentle liquid circulation that maintains cells in suspension. In conventional stirred tanks or bubble columns, the energy required for the mixing of the culture medium is introduced, at the one point in the reactor, via the stirrer or sparger, respectively. This results in a wide variation of shear forces with the shear being greatest near the stirrer (or the gas sparger). The cells may therefore encounter contrasting environments; either low shear forces but undesirable gradients in temperature and substrate concentration, or very high shear forces with sufficient heat and mass transfer. Either of these situations may have a detrimental effect on the fragile cells. The homogeneity of the shear distribution is the main advantage offered by airlift bioreactors, and this can be an important rationale for selecting this type of culture vessel for insect cell cultivation. Cell damage due to direct sparging can be minimized by the optimization of culture medium formulation and bubble generation.

Some researchers (Tramper et al., 1986; Agathos, Jeong and Venkat, 1990) have reported unsuccessful attempts at growing insect cells in small-scale airlift bioreactors, but recently published results (Maiorella et al., 1988; Weiss et al., 1988) on large-scale cultivation of insect cells in this type of reactor are quite promising.

Weiss et al. (1988) used commercially available, airlift bioreactors (LH Fermentation, Hayward, CA) of volumes 5, 10 and 40 liters, for cultivation of Sf-9 cell lines in serum-free medium. The sparged gases initially were N_2 air, and CO_2, at a 10% partial pressure of oxygen. On day 4 post inoculation, the oxygen partial pressure was increased to 20%. In the 5 liter airlift bioreactor, the aeration rate was 120 ml/min. In the 10 liter, the rate of sparging was initially 200 ml/min and was increased to 600 ml/min during the course of the experiment. In the 40 liter bioreactor, the sparge rate started at 400 ml/min, and as the cell growth increased, it was increased gradually to 800 ml/min. All bioreactors were inoculated with 2×10^5 cells/ml. The culture growing in a 5 liter vessel achieved a peak density of 6.4×10^6 cells/ml on day 8. This is comparable to the results obtained from small shaker flask cultures. The total cell densities obtained for the 10 liter and 40 liter bioreactors were 2.9×10^6 to 3.0×10^6 cells/ml. A rapid decrease of cell viability after 8 to 11 days of growth was observed.

A considerable contribution to the large-scale insect cell cultivation aimed at recombinant protein production has been reported by Maiorella et al. (1988). Their 2 liter working volume airlift fermentor (Chemap, South Plainfield, NJ) was successfully used to grow Sf-9 cells in serum-free medium. To protect cells from bubble damage and to limit foam formation, Pluronic polyol F-68 (0.1% w/v) was found to be effective. By choice of sparger orifice diameter, they were able to control the air bubble diameter at 0.5–1.0 cm. Over this size range, the bubbles do not interact to coalesce and, unlike smaller bubbles, have lower surface tension (the bubbles in this size range do not behave as rigid spheres). These factors minimized damage to cells during cell-bubble interactions. A gas (comprised of nitrogen, air, and oxygen) sparging rate of 420 to 1260 ml/min was sufficient to satisfy the oxygen demand without causing excessive turbulence or foaming. Under these conditions, the culture grew from 0.1 to 5×10^6 cells/ml over 7.5 days. This growth was comparable to that obtained in 100 ml spinner flasks.

Maiorella et al. (1988) have given the results of their oxygen transfer capacity measurements for traditional nonsparged vessels and airlift reactors. It was found that the oxygen transfer per unit volume of a nonsparged spinner flask decreases with increased vessel volume. The predicted value of 0.7×10^6 cells/ml, which can be supported in a 3 liter spinner flask without oxygen limitation, was in good agreement with experimental observations. The oxygen transfer capacity of an airlift bioreactor, on the other hand, increases as the vessel volume is increased (Lambert et al., 1987), and can be improved by controlling the airflow stream and its composition.

Schügerl (1990) compared the performances of stirred tank and airlift bioreactors for the production of secondary metabolites and found that, in the latter, the efficiency of oxygen transfer and specific production with regard to power input, substrate, and oxygen consumption were considerably higher than in stirred tank reactors, although the stirred tanks achieved higher cell concentrations and volumetric productivity. The most important aspects of airlift bioreactors design are discussed by Chisti in his recently published book (Chisti, 1989).

As mentioned previously, the addition of some polymer agents to the culture media proved to protect the insect cells against shear-induced damage due to agitation or sparging, or both (Tramper et al., 1986; Goldblum et al., 1990; Maiorella, et al., 1988; 1991). Murhammer and Goochee (1988) demonstrated the protective effects of Pluronic F-68 both in a 3 liter agitated, sparged bioreactor and in a 570 ml airlift bioreactor, for cultivation of Sf-9 cells and for production of recombinant β-galactosidase using an AcNPV baculovirus expression vector.

Recently, an extensive study on the mechanism of cell damage and Pluronic F-68 protection was presented by these researchers (Murhammer and Goochee, 1990b). The growth of Sf-9 insect cells was compared in two airlift bioreactors with similar geometry but different sparger designs. To one of these bioreactors, made by Ventrex Lab. (Portland, ME) the gas was introduced through a thin membrane distributor (pore size < 0.2 μm). In the second, a Kontes airlift bioreactor (Kontes, Vineland, NJ), two different porous stainless steel gas distributors were used, with average pore sizes of 5 and 20 μm. Medium supplemented with 0.2% (w/v) Pluronic F-68 provided protection to cells in the Ventrex airlift bioreactor and no protection to cells in the Kontes bioreactor when gas flow rates were the same in the two bioreactors. The results obtained suggested that cell damage can occur in the vicinity of the gas distributor. The damage at the sparger appeared to be a function of the gas flow rate, pressure drop and bubble size, the latter two variables being functions of the sparger design. The pressure drops across two different gas distributors in the Kontes bioreactor were considerably larger than in the Ventrex bioreactor. A model of cell death in sparged biorectors was presented (Murhammer and Goochee, 1990b).

Therefore, it seems that substantial progress has been made towards large-scale insect cell cultivation. Different types of bioreactor designs with direct or indirect air sparging were tested to grow insect cells. Promising results, especially using airlift bioreactors, were achieved.

3.4 BIOREACTORS WITH ATTACHED AND IMMOBILIZED CELLS

Cell density is a key factor affecting overall productivity, which is a classical parameter used to compare the efficiency of different types of bioreactors. The increased market for products that can be made from animal cells in culture has induced growing interest in high cell density bioreactors for large-scale cultivation of mammalian cells (Tyo and Spier, 1987). A low cell density obtained in conventional suspension cultures requires not only that production vessels be large to obtain a certain product capacity, but also makes downstream product recovery and purification more difficult, as desired product concentrations are also consistently low. The enhancement of cell densities in insect cell culture have, to some extent, been addressed.

The first attempts to achieve high cell densities were made with attachment-dependent cell lines used for large-scale production of baculoviruses (Weiss and Vaughn, 1986). Weiss et al., (1981) developed a method for growing a cell line from *S. frugiperda* in stationary large-scale culture. They used plastic roller bottles with growth surfaces of 850 cm^2 (corresponding to a medium volume of 150 ml), rotating at 1 revolution every 8.5 min. Exceeding this speed resulted in aggregation and uneven attachment of the cells. Average cell yield per roller bottle was 4.8×10^8 cells (3.2×10^6 cells/ml of liquid) and some of the highest specific growth rates (doubling time of 8.35 hr) for the *S. frugiperda* cell line cultures were obtained. Roller bottles, however, require extensive handling, labor and medium to produce high quantities of cells. Additionally, the scale-up of roller bottles and similar surface-dependent cell growth reactors is limited.

One method to increase surface to volume ratio is through the use of microcarriers. In contrast to roller bottles, this technique is very amenable to system monitoring and process control. Lazar et al. (1987) used cellulose-based microgranular microcarriers for the growth of *Aedes aegypti* (yellow fever mosquito) cells and scaled up the process to a volume of 8 liters. In microcarrier suspension systems, however, the problems of shear, oxygen transfer and heterogeneity exist. Animal cells in microcarriers are especially sensitive to damage from excessive agitation (Croughan, Hamel and Wang, 1987). Agglomeration of microcarriers is a potential problem.

Another bioreactor system for growth of an attached variant of *Trichoplusia ni* was successfully applied by Schuler et al., (1990). The insect cells were grown in a packed bed reactor packed with 3 mm nonporous glass beads. A separate unit was used for oxygenation of medium that circulated through the biocatalyst bed. Excellent growth of cells, with a doubling time of 16 hr, was obtained using the aeration reservoir as an airlift pump. In this design, both oxygen supply and shear are not strongly dependent upon bioreactor size, due to decoupling of the aerator from the main reactor. This can facilitate scale-up.

High cell densities and thus high bioreactor productivities may be achieved in immobilized cell systems. Agathos (1988) reported the cultivation of *Aedes albopictus* cells in immobilized culture using collagen based microspheres. A maximum cell density of 2×10^7/ml of bead was obtained. Given the fact that the beads were added at 40%(v/v) in the medium, much higher cell densities per bioreactor volume are possible. Taking into account a very low concentration of free cells in medium, the collagen-based beads demonstrated superior absorption or entrapment characteristics.

A novel technique recently applied to enhance cell density of insect cell culture is microencapsulation. Microencapsulation is now routinely used for the manufacture of monoclonal antibodies from hybridoma cells (Posillico, 1986). The cells are encapsulated in a semipermeable membrane and can be cultivated in standard suspension culture vessels. The advantages of this system, in addition to enhanced cell density, are that cells can be separated from the medium by gravity settling, as opposed to centrifugation, and the product may be either retained within the capsule (high molecular weight products) or diffused out from the capsule (low molecular weight products). The recent results by King et al. (1988) indicate attaining 8×10^7 cells/ml of capsule for *S. frugiperda* cells infected with a temperature-sensitive AcNPV baculovirus is possible. A disadvantage of this system is, of course, the cost of encapsulation and the possible oxygen transfer limitation in large-scale culture. Additionally, biocompatibility problems may arise (Smith et al., 1989). High density insect cell culture has shown to be a promising improvement in bioreactor performance. Several challenges, such as oxygen transfer limitation, cost of entrapment, etc., remain to be resolved before high cell density systems can be used in large-scale production.

CONCLUDING REMARKS

Due to potential applications for the production of valuable bioproducts, interest in insect cell culture has expanded considerably. A natural tendency exists to use the bioreactor technology currently applied to microbial and mammalian cells for insect cell culture. However, the shear sensitivity and, at the same time, high oxygen demand of insect cells are serious constraints on bioreactor design for cultivation of these cells. Many efforts have been made to adapt insect cells to grow in suspension in different types of bioreactors. Promising results in large-scale culture were achieved using airlift bioreactors. Low shear fields, good mixing, and simplicity are important advantages of this configuration. The concept of high cell density cultures must be also taken into consideration. These systems offer great potential for large scale production with significant improvements in economy over conventional systems. In this context, a better understanding of the biological implications of the effects of close cell packing is also needed.

REFERENCES

Agathos, S. N. (1988). Proc. Eng. Found. Conf. on Cell Cult. Eng., Palm Coast, Florida, paper number P.10.

Agathos, S. N., Jeong, Y. H. and Venkat, K. (1990). *Ann. N.Y. Acad. Sci., 589*:372–398.

Burges, H. D. (1981). *Microbial Control of Pests and Plant Diseases 1970–1980,* Academic Press, New York/London, p.949.

Chisti, M. Y. (1989). *Airlift Bioreactors,* Elsevier Applied Science, London/New York.

Croughan, M. S., Hamel, J. F., and Wang, D. I. C. (1987). *Biotechnol. Bioeng, 29*:130–141.

Eberhard, U. and Schügerl, K. (1987). *Develop. Biol. Standard., 66*:325–330.

Fleischaker, R. J. and Sinskey, A. J. (1981). *Eur. J.Appl. Microbiol. Biotechnol., 12*:193–197.

Glacken, M. W., Fleischaker, R. J. and Sinskey, A. J. (1983). *Ann. N.Y. Acad. Sci., 413*:355–372.

Goldblum, S., Bae, Y-K., Hink, W. F. and Chalmers, J. (1990). *Biotechnol. Prog., 6*:383–390.

Handa-Corrigan, A., Emery, A. N. and Spier, R. E. (1989). *Enzyme Microb. Technol., 11*:230–235.

Hink, W. F. and Strauss, E. M. (1976). *Invertebrate Tissue Culture—Application in Medicine, Biology and Agriculture* (E. Kurstak and K. Maramorosch, eds.), Academic Press, New York, pp.297–300.

Hink, W. F. and Strauss, E. M. (1980). *Invertebrate Systems in Vitro* (E. Kurstak, K. Maramorosch and A. Dübendorfer, eds.), Elsevier/North Holland, Amsterdam/New York, pp.27–33.

Hink, W. F. (1982). *Microbiol and Viral Pesticides* (E. Kurstak, ed.), Marcel Dekker, Inc., New York/Basel, pp.493–506.

Hu, S-L., Kosowski, S. G. and Schaaf, K. F. (1987). *J. Virol., 61:*3617–3620.

Karrer, D. (1988). Proc. Eng. Found. Conf. on Cell Cult. Eng., Palm Coast, Florida, paper number B4.

Khachatourians, G. G. (1986). *Trends Biotechnol., 4:*120–124.

King, G. A., Daugulis, A. J., Faulkner, P., Bayly, D. and Goosen, M. F. A. (1988). *Biotechnol. Lett., 10:*683–688.

Kunas, K. T. and Papoutsakis, E. T. (1990). *Biotechnol. Bioeng., 36:*476–483.

Kurstak, E. (ed.), (1982). *Microbial and Viral Pesticides,* Marcel Dekker AG Publishers, Basel.

Lambert, K. J., Boraston, R., Thompson, P. W. and Birch J. R. (1987). *Dev. Indust. Microbiol., 27:*101–106.

Lazar, A., Silberstein, L., Reuveny, S., and Mizrahi, A. (1987). *Develop. Biol. Standard., 66:*315–323.

Lengyel, J., Spradling, A., and Penman, S. (1975). *Methods in Cell Biology,* Vol.10 (D. M. Prescott, ed.), Academic Press, New York, pp.195–208.

Luckow, V. A. and Summers, M. D. (1988). *Bio/Technology, 6:*47–55.

Madisen, L., Travis, B., Hu, S-L. and Purchio, A. F. (1987). *Virology, 158:*248–250.

Maiorella, B., Inlow, D., Shauger, A. and Harano, D. (1988). *Bio/Technology, 6:*1406–1410.

Maiorella, B., Dorin, G., Carion, A. and Harano, D. (1991). *Biotechnol. Bioeng.,* 37:121–126.

Martignoni, M. E. (1984). *Chemical and Biological Controls in Forestry* (W. Y. Garner and J. Harvey, eds.), Amer. Chem. Soc., Washington, pp.55–67.

Merchuk, J. C. (1990). *Trends Biotechnol., 8:*66–71.

Miltenburger, H. G. and David, P. (1980). *Develop. Biol. Standard., 46:*183–186.

Murhammer, D. W. and Goochee, C. F. (1988). *Bio/Technology, 6:*1411–1416.

Murhammer, D. W. and Goochee, C. F. (1990a). *Biotechnol. Prog., 6:*142–148.

Murhammer, D. W. and Goochee, C. F. (1990b). *Biotechnol.Prog., 6:*391–397.

Papoutsakis, E. T. (1991). *Trends Biotechnol., 9:*316–324.

Podgwaite, J. D. (1985). *Viral Insecticides for Biological Control* (K. Maramorosch and K. E. Sherman, eds.), Academic Press, New York, pp.775–797.

Pollard, R. and Khosrovi, B. (1978). *Process Biochem., 13:*31–37.

Posillico, E. G. (1986). *Bio/Technology, 4:*114–117.

ProBioTech (1990). *Suppl. Process Biochem., 25* p.vii.

Prokop, A. and Rosenberg, M. Z. (1989). *Adv. Biochem. Eng., 39:*29–71.

Schügerl, K. (1990). *J. Biotechnol.*, *13*:251–256.

Shapiro, M. (1986). *The Biology of Baculoviruses, Vol. II* (R. R. Granados and B. A. Federici, eds.), CRC Press, Boca Raton, Florida, pp.31–61.

Shieh, T. R. and Bohmfalk, G. T. (1980). *Biotechnol. Bioeng.*, *22*:1357–1375.

Shuler, M. L., Cho, R., Wickham, T., Ogonah, O., Kool, M., Hammer, D. A., Granados, R. R. and Wood, H. A. (1990). *Ann. N.Y. Acad. Sci.*, *589*:399–422.

Smith, N. A., Goosen, M. F. A., King, G. A., Faulkner, P. and Daugulis, A. J. (1989). *Biotechnol. Tech.*, *3*:61–66.

Stockdale, K. and Gardiner, G. R. (1976). *Invertebrate Tissue Culture— Applications in Medicine, Biology and Agriculture* (E. Kurstak and K. Maramorosch, eds.), Academic Press, New York, p.267.

Streett, D. A. and Hink, W. F. (1978). *J. Invertebr. Path.*, *32*:112–113.

Tramper, J. and Vlak, J. M. (1986). *Ann. N.Y. Acad. Sci.*, *469*:279–288.

Tramper, J., Williams, J. B., Joustra, D. and Vlak, J. M. (1986). *Enzyme Microb. Technol.*, *8*:33–36.

Tramper, J., Smit, D., Straatman, J. and Vlak, J. M. (1988). *Bioprocess Eng.*, *3*:37–41.

Tramper, J., van den End, E. J., de Gooijer, C. D., Kompier, R., van Lier, F. L. J., Usmany, M. and Vlak, J. M. (1990). *Ann. N.Y. Acad. Sci.*, *589*:423–430.

Tyo, M. A. and Spier, R. E. (1987). *Enzyme Microb Technol.*, *9*:514–520.

Van Brunt, J. (1987). *Bio/Technology*, *5*:1118–1121.

Vaughn, J. L. (1968). In: *Proc. 2nd Intern. Colloq. Invertebr. Tissue Cult.*, Tremezzo, Italy, 1967(A. Baselli, ed.), Istit. Lombardo, Parisa, pp.119–125.

Weiss, S. A., Smith, G. C., Kalter, S. S. and Vaughn, J. L. (1981). *In Vitro*, *17*:495–502.

Weiss, S. A., Orr, T., Smith, G. C., Kalter, S. S., Vaughn, J. L. and Dougherty, E. M. (1982). *Biotechnol. Bioeng.*, *24*:1145–1154.

Weiss, S. A. and Vaughn, J. L. (1986). *The Biology of Baculoviruses, Vol. II* (R. R. Granados and B. A. Federici, eds.), CRC Press, Boca Raton, Florida, pp.63–87.

Weiss, S. A., Belisle, B. W., Chiarello, R. H., DeGiovanni, A. M., Godwin, G. P., and Kohler, J. P. (1988). *Conference on Biotechnology Biological Pesticides and Novel Plant-Pest Resistance for Insect Pest Management* (D. W. Roberts and R. R. Granados, eds.), Cornell University Press, Ithaca, New York, pp. 22–30.

Wudtke, M. and Schügerl, K. (1987). *Modern Approaches to Animal Cell Technology* (R. E. Spier and J. B. Griffith, eds.), Butterworths, London, pp.297–315.

Zhang, S., Handa-Corrigan, A., and Spier R. E. (1992). *Biotechnol. Bioeng.*, *40*:252–259.

4

Insect Cell Immobilization

Mattheus F. A. Goosen   *Queen's University, Kingston, Ontario, Canada*

4.1 INTRODUCTION

Immobilized animal cell cultivation and the production of recombinant biologicals appear to be made for each other. Cells and cell products can be concentrated and at the same time the desired product can be separated from the culture medium. However, while the technology is relatively successful at the laboratory scale, it suffers from the same potential pitfalls as industrial scale mammalian cell culture. At the laboratory scale, the main concerns with hollow fibers and microcapsules, for instance, are cell viability, control of membrane molecular weight cut off, and mass transport. Cost and complexity are not major factors. At commercial scale, on the other hand, price, complexity, and reliability are at the forefront. In the case of microcapsules, for instance, the efficient large-scale generation of uniform, small diameter droplets must go hand in hand with inexpensive biocompatible polymers and a simple encapsulation technique.

Techniques for immobilizing viable cells, tissue, and bioactive agents within protective semipermeable microcapsules have been developed in numerous laboratories, with applications ranging from cell transplantation to biotransformation to cell cultivation. In this chapter we will start by reviewing the polymers that can be employed for cell immobilization with particular emphasis on polymer toxicity. Following this, the development of an insect cell encapsulation system in our laboratory will be addressed. The primary reason for doing this is that the author is most familiar with this technology. Some of the problems faced by our group may be similar to those that have been encountered by others in bead and hollow fiber animal cell immobilization. Studies will then be outlined on how mathematical modelling has been used to increase our fundamental understanding of the problem of protein diffusion and cell growth in immobilized cell systems. The chapter will end with a brief discussion of the problems encountered in commercializing cell immobilization technology.

4.2 CHOOSING A POLYMER FOR CELL IMMOBILIZATION

Cell Immobilization Techniques

A variety of systems can be employed for cell and enzyme immobilization. These include, for example, microcarriers (Shahar and Reuveny, 1987), gel entrapment (Guo et al., 1990), hollow fibers (Shukla et al., 1989), and encapsulation (Lim, 1982). In principle, these methods fall into one of two categories: entrapment in a three dimensional polymer network or capture behind a semipermeable membrane. Among these methods, microcarriers and gel entrapment are probably the most popular. However, one common problem with both is leakage of the immobilized biocatalyst. Hollow fibers provide a solution to this problem by separating the cells from convected medium by keeping them on the outside or inside of a tubular asymmetric membrane. Microencapsulation provides another suitable method for cell immobilization. Numerous encapsulation techniques have been developed; interfacial polymerization (Chang, 1964), phase separation (Green, 1955), interfacial precipitation (Sefton and Broughton, 1982; Sefton et al., 1990), and polyelectrolyte complexation (Lim and Sun, 1980). One of the pioneers in this field, Chang (1964) was the first to encapsulate erythrocyte hemolysate and urease in semipermeable nylon membranes. Semipermeable microcapsules containing microbial cells, enzymes and viable red blood cells also have been prepared by Mosbach (1984), Chang et al., (1966) Hackel et al., (1975) and Kierstan and Bucke (1977). More recent systems include the polyacrylate capsules of Sefton et al. (1987) and Aebischer (1990), the chitosan/alginate system of Rha (1985) and McKnight et al. (1988), the alginate-polylysine (PLL) polyethyleneimine (PEI) system of Lim and Sun (1980), and the alginate-PLL technique of Goosen et al. (1985).

A complete review of all the available cell immobilization systems is beyond the scope of this chapter. Instead, in this section we will critically review the available cell encapsulation techniques with particular emphasis on polymers for capsule formation and the problem of polymer toxicity.

Polyelectrolyte complexation has been known to be a mild process for cell encapsulation. Membrane formation occurs through an interaction between negatively and positively charged water soluble polymers. A variety of viable cells can be encapsulated by this method (Table 1). Many researchers (Lim, 1984; King et al., 1987, 1989; Yoshioka et al., 1990; Reilly et al., 1990) have shown that culture of encapsulated cells can provide a higher cell density and higher product concentration than suspension culture. It is noteworthy that the first industrial monoclonal antibody production technique employing microcapsules was demonstrated by Damon Biotech (now Abbott Biotech) (Posillico, 1986). They

Table 1 Encapsulation of Viable Cells by Polyelectrolyte Complexation

Membrane polymers	Cells encapsulated	References
Alginate-polylysine	Rat islets of Langerhans	Lim and Sun, 1980
		Sun and O'Shea, 1985
		Goosen et al., 1985
		McCarthy et al., 1988
	Hybridoma cells	Posillico, 1986
		King et al., 1987
		Lim, 1984
	Insect cells	King et al., 1989
	Hepatocytes	Lim, 1984
	Tumor cells	
	Andrenal cortical cells	
	Tumor infiltrating lympho-	Reilly et al. 1990
	cytes	
	Human T cells	
Alginate-polyacrylate	Human red blood cells	Gharapetian et al. 1986
Cross-linked albumin	Hybridoma cells	Tice and Meyers, 1985
Chitosan-CM-cellulose	Hybridoma cells	Shiotani et al. 1986
		Yoshioka et al. 1990
Chitosan-alginate	Microbial cells	Rha et al. 1988
	Plant cells	Knorr et al. 1987
Chitosan-k-carrageenan	Plant cells	Knorr et al. 1987

showed that both the concentration and purity of the intracapsular product were approximately 100 times greater than could be achieved in suspension culture (i.e., 30 gram antibody was harvested from 15 liters capsules; this amount is usually obtained from 500 liters medium by conventional culture).

Polymers for Cell Immobilization

Materials such as polyamides, polyacrylates, albumin, Eudragit, RL, alginate, CM-cellulose, k-carrageenan, and chitosan have been employed to form capsules. Among these, polymers such as alginate, CM-cellulose, K-carrageenan, and chitosan, which come from natural sources, are of particular interest because they are of low cost and naturally regenerable (Figure 1). From this group, alginate is one of the most popularly used gel matrix forming materials. This polysaccharide exists in the marine brown algae as an intracellular matrix gel and comprises up to 40% of the dry matter (Skjak-Braek, 1988). In the molecular sense, alginate is a collective term for a family of linear binary copolymers containing 1,4-linked beta D-mannuronic acid (M-block) and alpha-L-guluronic acid (G-block) residues in varying proportions and sequential arrangements. Alginate exhibits many different physical, chemical and biological properties, depending on its molecular structure. In terms of gel forming ability, structure plays an important role in membrane formation. It was found that alginate with a structure of G-block monomers over 20 units could selectively bind divalent cations to form inotropic gels with M-block monomers (Lapasin et al., 1988). In contrast, alginate with the sequence of G-residues and M-residues in strictly alternating manner did not form gels with calcium ions at all (Skjak-Braek, 1988).

Carrageenans constitute a broad class of sulfated linear galactans extracted from seaweed. The molecular structure involves an alternate 1,3-alpha-D- and 1,4-beta-D- linkage. In this polysaccharide family, iota- and kapp-carrageenans are gelling polymers, and mu- and lambda-carrageenans are non-gelling due to structural irregularities (kinks). It was reported that even one kink in 200 residues could dramatically reduce the gel strength (Yalpani, 1988). The gelation structure is presently described by a double helix model. Carboxymethyl (CM) cellulose is another polyanionic polymer, which can be derived from cellulose by reacting it with sodium monochloroacetate under alkali conditions (Iwata et al., 1985). This polymer has been widely used in food additives, cosmetics, detergents and basic pharmaceutical materials.

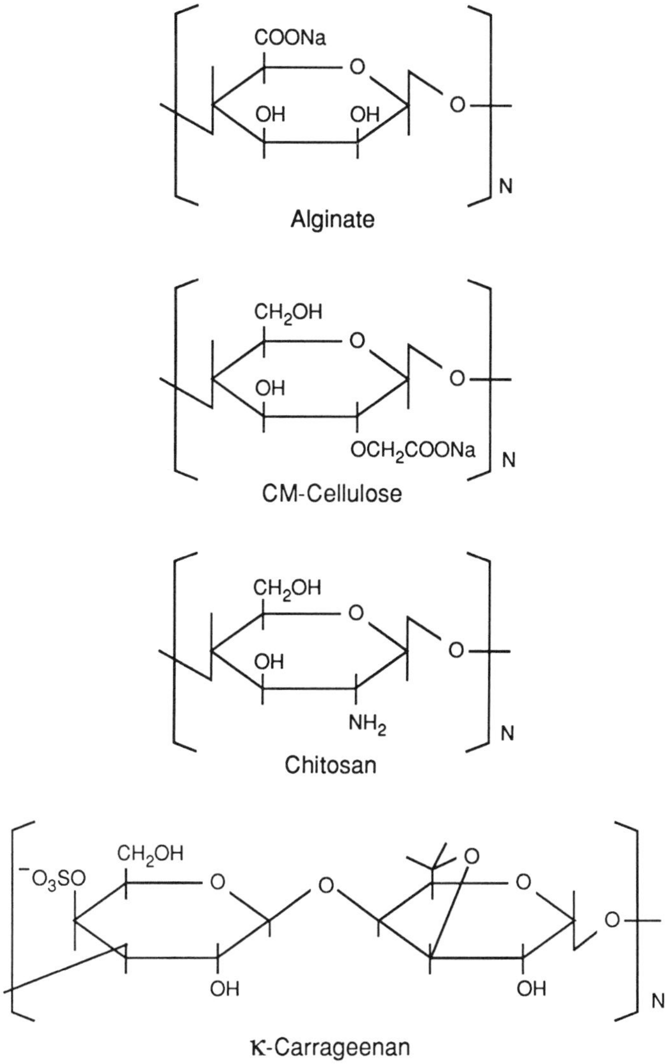

Figure 1 Chemical structure of polymers employed in animal cell immobilization.

In addition to polyanionic polymers, one important polycation is chitosan, a polysaccharide with 1,4 linked beta-2-amino-2-deoxy residues. This polymer is derived from deacetylated chitin, which is extremely abundant in nature, existing widely in mushrooms, yeasts, and the hard outer shells of insects and crustaceans. About 50% to 80% of the organic compounds in the shells of crustacea and the cuticles of insects consists of chitin (Muzzarelli, 1977). Early applications of chitosan were limited to such areas as surgical treatment for wound healing by the Koreans and Mexicans (Allan et al., 1984), and as additives for painting (Alper, 1984). More recent applications of chitosan include use as a powder for removing cholesterol and fats from food, as a biodegradable suture for surgeons, as a coating for preserving freshness of fruits, and as a filter for removing harmful metals and elements from polluted water (Strauss, 1988).

Chitosan has been shown to be an outstanding polymer for membrane formation due to high density amino groups in the molecule. In 1978, Hirano showed that N-acetyl chitosan membranes were ideal for controlled agrochemical release. In 1984, Rha et al. employed chitosan to form capsules for immobilization of living cells. The capsules were formed in a single step by covering a liquid alginate core with a complexed chitosan-alginate membrane. Since then, the encapsulation of other cells, such as hybridoma and plant cells, in chitosan capsules has been widely investigated (Knorr et al., 1987; Shiotani et al., 1986; Yoshioka et al., 1990).

Apparent Polymer Toxicity

In order to entrap animal cells in polymer capsules, two requirements must be satisfied. First, the polymers, which come into contact with the cells, must be nontoxic. Secondly, mass transport conditions in the microcapsules must be compatible with the cells, which means that in order to meet the demands for cell growth and cell maintenance there must be no significant mass transfer hindrance. A variety of nutrients, including important growth factors and proteins, should be able to freely diffuse from the bulk medium across the capsule membrane, through the polymer solution/matrix, to the cells.

In work performed in our laboratory, it was found that one of the natural membrane forming polymers, chitosan, appeared to be toxic to insect cells (Smith et al., 1989). Results showed that a one minute contact between insect cells and a 0.1% chitosan solution could result in an 80% loss of cell viability (Table 2). Alginate, on the other hand, ap-

Table 2 Toxic Effect of Various Chitosans on Insect Cells

Type of chitosan*	Approximate Mv	Percentage of viable cells**
SeaCure+ (water-soluble)	405,000	49 ± 32
Protasan	968,000	20 ± 6
SeaCure-F:		
Low molecular weight	1,000,000	46 ± 14
Medium molecular weight	1,247,000	46 ± 24
Extra high molecular weight	1,641,000	50 ± 11

*Protan
**Contact time = 1 minute
Source: Smith et al., 1989.

Table 3 The Effect of Alginate on Cell Viability with Varying Alginate/Cell Contact Time and Alginate Concentration

Alginate concentration	Contact time (min)	Percentage of viable cells
1.4% (w/v)	1	80 ± 10
	5	78 ± 24
	20	96 ± 7
0.7% (w/v)	1	95 ± 5
	5	75 ± 6
	20	80 ± 1
KCl Control	1	100
	5	100
	20	100

Source: Smith et al., 1989.

peared to be quite acceptable (Table 3). However, the results with chitosan may not be the final word, since Rha et al. (1987) apparently were successful in growing hybridoma cells in chitosan capsules.

Even if all membrane forming polymers in the encapsulation process are nontoxic to living cells, conditions inside the microcapsule may still be incompatible with the cells. King et al., (1989), for example, found that the intracapsular alginate concentration in alginate-poly-l-lysine capsules was a key factor in determining insect cell survival and growth. When the concentration of alginate employed in the encapsulation procedure was over 0.75%, cell growth inhibition was observed (Table 4). The results suggested that increased mass transfer resistance in the alginate

Table 4 Biocompatibility Tests: Cells Suspended in Alginate Gel

Alginate concentration	Cell viability
1.5% (w.v)	All cells enlarged, and dark, most dead.
1.3% (w/v)	All cells enlarged, and dark, most dead.
1.0% (w/v)	Some cells dark and granular while some are round and healthy. Some cells have multiplied to form small masses.
0.75% (w/v)	Good cell growth. Many cell masses.
0.50% (w/v)	Good cell growth. Many cell masses.

Source: King et al., 1989.

core to nutrient diffusion might have been the primary reason for cell death at high alginate concentrations.

The effective diffusion coefficient of many substances, such as glucose (Itamunoala, 1987; Chresand et al., 1988), lactic acid (Chresand et al., 1988), and proteins (Tanaka et al., 1984; Martinsen et al., 1989), through gel matrices has been measured. For small molecules (i.e., molecular weight below one hundred), the diffusion coefficients in gels usually range from 50% to 100% of those in water (Guiseley, 1989). The diffusion of macromolecules in gel matrices is rather complex. Molecular weight and configuration are two important factors influencing the diffusion rate (Martinsen et al., 1989). Usually, the diffusion rate decreases with an increase in molecular weight (Guiseley, 1989). When the molecular weight of the protein is over 150,000, its diffusivity is about 16% to 50% of that in water. Tanaka et al. (1984) interestingly found that large molecules with molecular weights over 65,000 could not diffuse into alginate gels, but they could diffuse out. Yuet et al. (1992) studied the diffusion of two low molecular weight nutrients, glucose and glutamine, through sodium alginate solution. He found that the viscosity of the solution had a remarkable influence on the diffusivity of glucose. When the viscosity was increased from 20 to 50 cSt, the diffusivity of glucose was reduced to 8% of that in water. In 1954, Cook and Smith showed that as the polymer size increased, the diffusion rate of macromolecules became much more dependent on polymer concentration. Yuet's results suggest that a high polymer concentration, which usually results in a high viscosity, may play a critical role in hindering nutrient mass transfer. Studies by others also support the speculation that there is a barrier effect with regard to nutrient mass transfer in polymer solution. In particular, published reports on cell growth in microcapsules showed that the cells appeared to grow preferentially near the interior membrane surface of capsules (Posillico 1986; King et al., 1987). More recently, Kwok et al., (1990) mathematically modeled the diffusion process. They pointed out that an impermeable core apparently existed in alginate-polylysine capsules, its effective size being proportional to the molecular weight of the diffusing protein. This will be covered in more detail later in this chapter.

4.3 THE SEARCH FOR A SUITABLE INSECT CELL ENCAPSULATION TECHNIQUE

In 1986, Posillico reported for the first time the use of microencapsulated animal cells for the production of monoclonal antibodies in multigram quantities. They reported, however, that their cells appeared to grow preferentially near the interior surface of the microcapsule membrane, and speculated that this could have been due to mass transfer limitations during the cell culture, or to the presence of a viscous intracapsular alginate solution. (This, in hindsight, was actually a forewarning of the troubles that we were to run into in our efforts to develop an insect cell encapsulation system.) At about the same time experiments in our laboratory showed that the membrane molecular weight cutoff of alginate-poly-l-lysine (PLL) microcapsules could be controlled from 60×10^3 to $\sim 300 \times 10^3$ by varying the molecular weight of the PLL, the alginate-PLL reaction time, and the PLL concentration. These findings led to the development of a patented multiple membrane microcapsule system which had a reduced intracapsular alginate concentration. (Goosen et al., 1990) (Figure 2). In comparison to the conventional single membrane system, higher intracapsular cell densities and product concentrations ($\sim 300\%$) were obtained (King et al., 1987) in about half the cell culture period. In addition, there appeared to be better retention of the cell product by the multiple membrane microcapsule.

While most of the previous work had been done with hybridoma cells, the main objective of our more current research was to investigate whether insect cells (*Spodoptera frugiperda*) infected with a temperature-sensitive mutant of the *Autographa californica* nuclear polyhedrosis virus (AcNPV ts-10) could be cultured to a high cell density in microcapsules, and whether the virus and recombinant product could be successfully grown and concentrated within the capsules. Surprisingly, even though hybridoma cells grew well in capsules, insect cells could not be cultured in either single or multiple membrane capsules. As a result of this, we spent

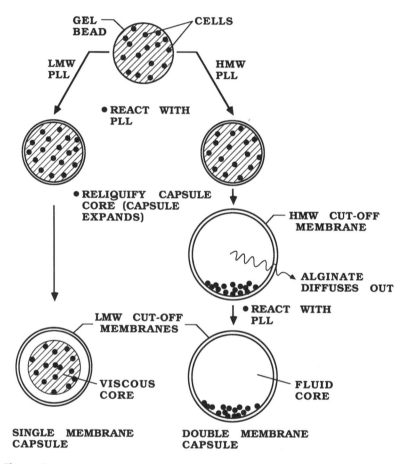

Figure 2   A comparison of cell encapsulation techniques.

about six months going through a systematic series of studies before we found a solution to our problem.

Biocompatibility Tests

Biocompatibility tests were performed on the encapsulation solution and the polymers (Smith et al., 1989). The encapsulation solutions, CHES, $CaCl_2$, KC1, and sodium citrate, for example, did not have any apparent effect on the growth of the insect cells. However, while direct exposure of the cells to low molecular weight poly-l-lysine (Mv = 22,000), resulted in virtually no loss of cell viability, exposure of the cells to higher molecular weight poly-l-lysine (Mv = 102,000 and Mv = 270,000) resulted in a 75% loss of cell viability. In a different set of experiments, King et al. (1987) demonstrated that a direct relationship exists between the viscosity average molecular weight of the poly-l-lysine and the membrane molecular weight cut off of a microcapsule. Comparing these two different yet related studies, we concluded that due to the loss in cell viability with higher molecular weight poly-l-lysine, care must be taken in insect cell microencapsulation when attempting to increase the capsule membrane molecular weight cutoff.

The results of the toxicity tests involving various molecular weight chitosans (an inexpensive alternative to PLL) are summarised in Table 2. Protasan, an ultrapure medical grade chitosan, had the greatest toxic effect on the cells, resulting in a decrease in cell viability of approximately 80%. The unpurified chitosans of different molecular weights did not appear to have as significant an effect on cell viability (greater than 50% viable). We can speculate that this toxic effect of the ultrapure chitosan may have been due to the introduction of contaminants during the purification process. Another possibility is that the cruder chitosans contain other material (i.e., proteins) that protect the cells, and that this material was removed during the purification process. The unpurified chitosans tested had viscosity average molecular weights in the range of 400,000 to 1,641,000. In contrast to poly-l-lysine, the molecular weight of the chitosan did not seem to have an effect on cell viability.

The alginate concentration used most often in mammalian cell encapsulation is approximately 1.4% (w/v). Employing the cell-polymer contact test, this alginate concentration did not have a significant deleterious effect on insect cell viability (Table 3). Increasing the alginate-cell contact time had no apparent effect on cell viability; thus, prolonged exposure of insect cells to alginate during the formation of microcapsules should not have affected cell viability.

In contrast, an alternative biocompatibility test not only gave us different results, but also a solution to our problem. In this test a layer of insect cells was allowed to grow and adhere on the bottom of a T flask, then the medium was decanted and replaced with sodium alginate solution. Fresh medium was then poured on top of the alginate layer. The alginate layer subsequently gelled. The flask was then placed in an incubator. We found that only at alginate concentrations of 0.75% or less was cell growth observed (Table 4). With this test, cell density could not be measured directly since the cells were immobilized at the bottom of an alginate gel layer. These results suggested that insect cell growth inhibition was probably caused by a mass transport limitation. The nutrients may not have been able to diffuse through the alginate when its concentration was above 1%.

Cell Culture Studies

The observation that the alginate concentration might be a key factor in the growth of the cells was confirmed with immobilized cell culture studies. Insect cells encapsulated using a conventional single membrane capsule (molecular weight cutoff of $\sim 6 \times 10^4$) with an initial intracapsular alginate concentration of 1.4%, grew poorly. In this case, the membrane was relatively impermeable and thus only a small fraction of the alginate (molecular weight 3.5×10^5) could diffuse through. Cells encapsulated in single membrane capsules with molecular weight cutoff of $\sim 3 \times 10^5$ grew much better, though in isolated clumps, presumably because there was increased mass transfer and because more of the intracapsular alginate could diffuse out. Similar clumped cell growth has been observed by other workers who cultured hybridoma cells in single membrane microcapsules (Rupp, 1985). Good cell growth was still not achieved by our group when the two step (multiple membrane) encapsulation procedure was employed (Figure 2). However, our problem was finally solved when the initial alginate concentration was reduced from 1.4% to 0.75% in the multiple membrane encapsulation procedure. In this case we observed that encapsulated cells grew well with the growth rate comparable to that of suspended cells (Table 5). The good cell and viral growth observed in these improved capsules was presumably due to

Table 5 Comparison of Growth Rates of Encapsulated and Suspended Insect Cells

Culture system	Specific growth rate (day^{-1})	Doubling time (h)
Suspended cells		
Shaker flasks	0.66	25
Monolayer (T-Flasks)	0.78	21
Immobilized cells		
Microcapsules	0.55	30

Source: King et al., 1989.

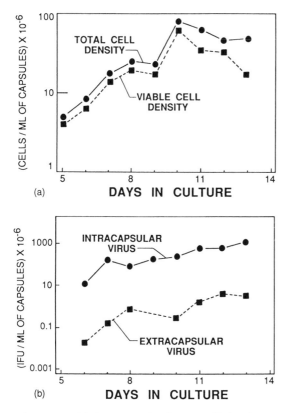

(a)

(b)

Figure 3 (a) Intracapsular insect cell density as a function of time. Batch culture consisted of 1 ml capsules in 30 ml medium. (b) Intracapsular and extracapsular virus concentration as a function of culture time. At day 5, the culture temperature was dropped from 33 to 27°C to initiate virus growth in the infected insect cells. MOI = 0.02 (King et al., 1989).

the lower intracapsular alginate content (Figures 3 and 4). It is unclear at this time whether we have two distinct alginate-PLL membranes or one interacting membrane. The term multiple membrane refers more to the method of capsule preparation rather than to the final membrane. On the negative side, we observed that capsules with a reduced alginate content had a greater tendency to collapse. We can speculate that the presence of a viscous alginate core provided physical support for the capsule membrane. Once the concentration of alginate in the core was lowered, the level of physical support keeping the capsule membrane expanded decreased.

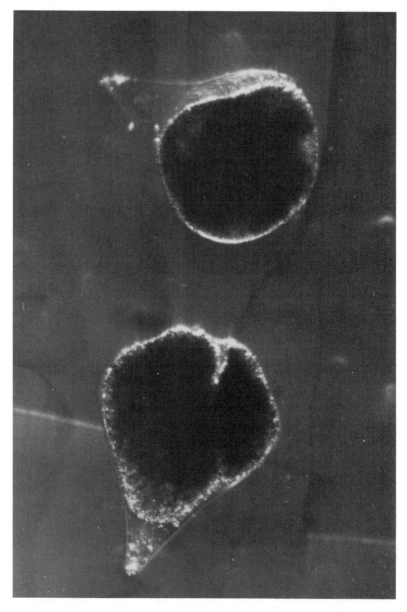

Figure 4 Culture of insect cells in multiple-membrane microcapsules at day 10. Initial sodium alginate concentration is 0.7% (w/v).

In summary, these studies added further documented evidence that the polymer core, not the capsule membrane, provides the main barrier for effective immobilized cell culture. In addition it was shown that the ability to control virus replication by lowering the culture temperature has allowed for growth and concentration of virus inside of cell filled microcapsules. Recent work (King et al., 1990, unpublished) has shown that it is possible to obtain intracapsular recombinant protein (chloramphenicol acetyl transferase) concentrations in excess of 20,000 units/ml (4 g/L).

4.4 UNDERSTANDING THE MASS TRANSPORT PROBLEM BY MODELING PROTEIN DIFFUSION AND CELL GROWTH

While it is apparent that there has been considerable development and application of cell encapsulation technology, a fundamental understanding of the mass transport behavior of proteins, nutrients, and waste products diffusing into and out of the capsules appears to be lacking. The characterization of this behavior is important since cell viabilities are affected by the transport of nutrients and oxygen into a capsule and may be affected by product diffusion out of the capsule. Furthermore, an understanding of cell growth in microcapsules is necessary to analyze, operate and optimize the process more efficiently.

Although numerous studies on the diffusion characteristics of the microcapsule system have been conducted, (King et al., 1989), there have been only a few attempts to address mathematical modeling of microcapsules. Mogensen and Vieth (1973) studied the mass transfer properties of enzymes in semipermeable microcapsules. Heath and Belfort (1987) also provided a mathematical treatment of microcapsules entrapping biocatalysts. However, no cell growth was taken into account in either paper. The primary objective of this section is to go over the development of a general diffusion model for predicting mass transport phenomena in microcapsules, by comparing calculated and experimental diffusion profiles, using Fickian diffusion equations (Kwok et al., 1990). The mathematical derivation for the mass transport phenomena and models in membrane permeation were developed following the analogies presented by Flynn et al., (1974) and Crank (1975). A mathematical model for animal cell growth in microcapsules will also be briefly outlined (Yuet et al., 1992).

Model Development

A fundamental solution for the diffusion of proteins from a bulk solution into a spherical capsule through a semipermeable membrane can be derived from the hollow sphere model reported by Carslaw and Jaeger (1959). Fickian diffusion is assumed. This can be justified by the fact that according to previous investigation, the water content of the alginate-PLL membrane is approximately 90% by weight. This means that it can be treated as a hydrogel or a grossly porous membrane whose pores are filled with water (Goosen et al., 1985). In a membrane of this nature, even though there may be fixed ionic groups present, these ionic groups may not exert a significant effect on determining the membrane permeability with respect to a particular group of diffusing molecules; rather, it is the size and shape of the molecule that will determine the permeability (Lakshminarayanaiah, 1969). In other words, a grossly porous membrane, even if it is charged, may not possess the semi-permeability brought about by the ionic groups in the membrane. The effect of material transport in the form of forced diffusion due to an electrostatic driving force in the membrane, therefore, can be assumed to be negligible when compared to that of ordinary Fickian diffusion. This assumption may be further justified by the fact that the membrane was treated with dilute sodium alginate after liquification of the core; the negatively charged carboxyl groups in alginate would presumably have neutralized any remaining positively charged amino groups in the PLL-alginate membrane.

Why model protein diffusion? The molecular weight cutoff of the capsule membrane was first characterized in our laboratory several years ago (Goosen et al, 1985) by employing proteins from molecular weight calibration kits. We thus had diffusion data readily available for our modeling studies.

The physical system consists of a viscous liquid or gel enclosed within a semipermeable membrane (Figure 5). Microcapsules, which are initially free of proteins, are suspended in a protein solution. With the existence of a concentration gradient, the protein molecules begin moving into the membrane and diffuse across it.

In developing the model the following assumptions have been made:

1. The extracapsular solution is well mixed, so the bulk protein concentration is uniform (no concentration gradients exist).
2. The diffusion of protein in the sodium alginate liquid shell or core is very rapid compared to the diffusion through the semipermeable membrane.
3. All capsules are of identical size, each capsule containing at any time

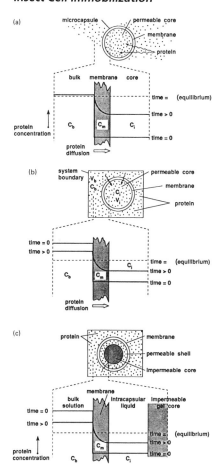

Figure 5 Protein diffusion models: Diffusion of protein (solute) from bulk solution into a spherical capsule. The protein concentration profiles assume well-mixed bulk and core solutions, no stagnant boundary layers and no discontinuities at the interfaces (Kwok et al., 1990). (a) Case 1: Infinite bulk solution. Entire capsule core is permeable to diffusing protein. (b) Case 2: Limited volume bulk solution. (c) Case 3: Limited volume bulk solution. A partially impermeable capsule core. The permeable fraction, for example, might be a liquid, while the impermeable fraction might be a gel.

the same amount of albumin. This assumption has been experimentally confirmed (Goosen et al., 1985).

4. The solubility of protein in the extracapsular solution is equal to its solubility in the alginate core (partition coefficient equal to 1). This assumption should be valid since the alginate core is a dilute solution of alginate in water (< 0.015 g/ml); thus the solubility of protein should be very similar to that in the extracapsular solution.

The governing diffusion characteristics in the membrane can be described by the equation:

$$\frac{\partial C_m}{\partial t} = \frac{1}{r^2} \frac{\partial}{\partial r} \left(D_{mp} r^2 \frac{\partial C_m}{\partial r} \right)$$

(1)

where $\partial C_m/\partial t$ denotes the rate of change in the concentration of protein (or solute) in the membrane, D_{mp} is the diffusivity of protein in the membrane, r is the radial position (direction of diffusion), $\partial C_m/\partial r$ is the protein concentration gradient in the membrane, and t is time. At the membrane/intracapsular solution interface negligible resistance is assumed [Figure 5(a)].

The intracapsular protein concentration increases as a result of protein diffusion through the membrane. If we assume that the capsule membrane offers the main resistance to protein diffusion, then a uniform protein concentration most likely exists in the alginate core (i.e., $D_{ip} >> D_{mp}$ where D_{ip} is the diffusivity of protein in the alginate core). The rate of change of protein in the alginate core, dMi/dt, can then be related to the protein concentration gradient at the membrane-core interface, according to Fick's first law of diffusion (Flynn et al., 1974):

$$\frac{dM_i}{dt} = A_i D_{mp} \left(\frac{\partial C_m}{\partial r} \right)_1$$

(2)

where A_i is the inside membrane surface area and D_{mp} is the diffusivity of protein in the membrane. Rearranging Equation (2), the following expression is obtained:

$$\frac{dC_i}{dt} = \frac{1}{V_i} \frac{dM_i}{dt} = \frac{A}{V_i} D_{mp} \left(\frac{\partial C_m}{\partial r} \right)_1$$

(3)

where V_i is the intracapsular volume available for protein to diffuse into, and C_i is the intracapsular protein concentration.

In the case of a limited volume, well-mixed bulk (extracapsular) solution, the bulk protein concentration, C_b, changes as the protein diffuses into the microcapsules [Figure 5(b)]. By applying Fick's law of diffusion, the rate of solute entering the membrane from the bulk solution can be expressed as

$$\frac{dM_b}{dt} = A_2 D_{mp} \left(\frac{\partial C_m}{\partial r} \right)_2 ; A_2 = NA_2$$

(4)

where dM_b/dt is the molar rate of change of protein in the bulk, N is the number of microcapsules, A_2 is the outside surface area of the microcapsule, and $(\partial C_m/\partial r)_2$ is the protein concentration gradient at the membrane-bulk solution interface. The rate of change in bulk solute (protein) concentration, (dC_b/dt), can be obtained by dividing the total change in amount of solute by the total bulk volume, V_b,

$$\frac{dC_b}{dt} = -V_b \frac{dM_b}{dt} = \frac{A_2}{V_b} D_{mp} \left(\frac{\partial C_m}{\partial r} \right)_2$$

(5)

The bulk solution is assumed to be well mixed. For spherical capsules, the total surface area, A_2, and total volume of microcapsules, V_c, is given by:

$$A_2 = N4\pi R^2 \text{ and } V_c = N\frac{4}{3}\pi R^3$$

where R is the capsule radius. Rearranging Equation (5), we have the bulk solute balance equation:

$$\frac{dC_b}{dt} = \frac{3}{R_2} \frac{V_c}{V_b} D_{mp} \left(\frac{\partial C_m}{\partial r} \right)_2$$

(6)

Equations (1), (3), and (6) provide the mathematical model for the protein mass balance in the membrane, intracapsular solution, and bulk solution respectively.

All of the previous equations for the intracapsular solution assumed a totally permeable core. Under certain conditions however, the possibility exists that the alginate core in the microcapsule may be impermeable to the diffusing protein (Goosen et al., 1985, Posillico, 1986). Tanaka et al. (1983) reported that calcium alginate gel was impermeable to large substrate molecules such as albumin (molecular weight 67,000), gamma globulin (molecular weight 150,000) and fibrinogen (molecular weight 330,000). Hence, it would be reasonable to assume that residual calcium alginate inside the capsules may reduce the effective volume available for diffusion of proteins [Figure 5(c)]. The inner (impermeable) alginate gel core in this model is assumed to be spherical and is located at the center of the capsule. The effective or free volume (i.e., the spherical shell between the membrane and the core), V_f, can be expressed in terms of a fraction, F of the total intracapsular volume V_i.

$$V_f = FV_i \qquad (7)$$

where F is a constant which satisfies the constraints: $0.0 \leq F \leq 1.0$. For a totally permeable core, $F = 1.0$ and for a totally impermeable core, $F = 0$.

The change in the intracapsular protein concentration, dC_i/dt, in the free volume, V_f, can therefore be expressed by replacing total intracapsular volume, V_i, with V_f, in the mass transport equation for diffusion of protein into a finite volume [Equation (3)]. The change in intracapsular protein concentration, C_i, within the free volume inside the capsule is:

$$\frac{dC_i}{dt} = \frac{A_i}{V_f} D_{mp} \left(\frac{\partial C_m}{\partial r} \right)_1 = \frac{A_1 D_{mp}}{FV_i} \left(\frac{\partial C_m}{\partial r} \right)_1 \qquad (8)$$

The fraction F can be evaluated experimentally, or an arbitrary value can be assigned to it for modeling purposes.

The intracapsular protein concentration, C_i, the bulk protein concentration, C_b, and the membrane protein concentration, C_m, can then be determined by the simultaneous solution of Equation (1) (protein diffusion through membrane), Equation 6 (bulk protein diffusion) and Equation (8) (protein diffusion into the capsule core).

A Comparison of Calculated and Experimental Diffusion Results

The intracapsular solute concentration was calculated as a function of time and membrane diffusivity, by solving the membrane mass transport equation and the equation for solute diffusion into a finite volume, Equation (3), simultaneously. The value of the membrane diffusivity was varied to obtain the best possible agreement with experimental data. At a diffusivity of 30×10^{-12} m²/min, the theoretical equilibrium between intracapsular and extracapsular protein was reached within two hours. In these calculations, for the case of a microcapsule with a liquid core, it was assumed that the capsule membrane was initially free of solute and that the concentration of the bulk protein solution stayed constant at C_b. The initial and boundary conditions were taken as:

$$C_m(R_1, t) = C_i$$
$$C_m(R_2, t) = C_b$$
$$C_m(r, 0) = 0$$

In contrast to the theoretical diffusion profiles, experimental studies on albumin diffusing into the capsules showed a marked difference. The increase in the apparent intracapsular albumin concentration in capsules made with PLL of molecular weight of 227,900 and 102,600 decreased significantly after one hour. A comparison of the theoretical and experimental diffusion results supported our earlier speculation by suggesting that the capsules may not have had a completely permeable core. In the experimental diffusion studies, the capsules, after being ruptured, were centrifuged, and sodium citrate was added to recover the intracapsular solution.

In several of the studies, the size of the microcapsules was measured throughout both encapsulation and diffusion studies. Large volume expansion was observed during the core liquification with sodium citrate (Kwok et al., 1990). The degree of expansion was found to depend on the molecular weight of the PLL employed in the encapsulation procedure. For capsules prepared with PLL of high molecular weight ($M_v = 227,900$), microcapsules expanded to more than one and a half times the original gel bead diameter after incubation in citrate solution. The diffusion of albumin into the capsules did not appear to affect the capsule size. Similar results were also observed by King et al., (1987). The assumption used in our mathematical analysis, of an absence of membrane swelling during solute diffusion, was therefore verified.

Determination of Membrane Diffusivity

The effects of PLL-alginate reaction time on the membrane molecular weight cutoff has been studied previously by King et al., (1987), who measured the decrease in extracapsular protein solution as the solute diffused into the capsule. They suggested that the reaction time apparently reduced membrane diffusivity by tightly crosslinking the two polymers (polyelectrolyte complexation). The results of their studies showed that at high reaction times (40 min) the membrane was completely impermeable to albumin, as reflected by the constant absorbance reading. At a lower reaction time, the membrane was permeable to albumin but equilibrated at different final absorbance values, suggesting the presence of a core which was partially impermeable to the diffusing protein.

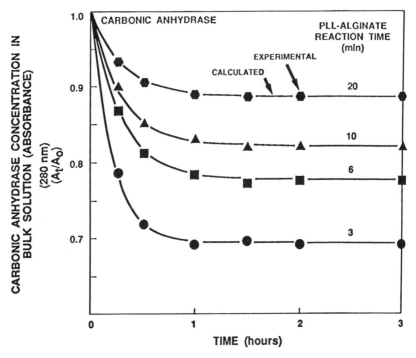

Figure 6 Comparison of calculated and experimental diffusion profiles with carbonic anhydrase ($M_v = 29,000$) as diffusing protein. The M_v of PLL was 21,000. (Kwok et al., 1990)

Based on the mathematical model derived for solute diffusion into a capsule of limited free volume [Equation (8)], the diffusivities of the capsule membranes were determined by superimposing theoretical curves on the experimental data previously obtained by King et al., (1987). In the case of albumin the membrane diffusivity, D, was determined from the initial slope of the curve, while the apparent size of the impermeable gel core was calculated from the final equilibrium protein concentration. The fit was excellent (Figure 6). It was found, for example, that the membrane was more permeable when a shorter PLL-alginate reaction time was employed (Tables 6 and 7). In the case of albumin, the membrane diffusivity, D, decreased from 16.3 to 2.0 × 10^{-12} m²/min (or 2.7 × 10^{-9} to 3.3 × 10^{-10} cm²/s) when the PLL-alginate reaction time increased from 3 to 20 min. These appear to be unusually low values compared to literature values for the diffusivity of albumin

Table 6 Calculated Membrane Diffusivity and Size of Impermeable Gel Core with Albumin as Diffusing Protein

PLL-alginate reaction time (mins)	Capsule* diameter (μm)	Apparent diameter of impermeable gel core** (μm)	Apparent volume fraction of intracapsular free space	Apparent membrane diffusivity** ($\times 10^{12}$)(m²/min)
3	1,000	838	0.393	16.3
6	1,000	909	0.227	6.7
10	1,000	958	0.094	2.8
20	1,000	973	0.051	2.00

M_v of PLL = 21,000
*Microcapsule diameter arbitrarily set at 1000 μm (R_1 = 500 μm).
**Values for gel core diameter and membrane diffusivity were calculated using Equations 1, 6 and 8, by fitting the calculated diffusion curve to the experimental results.
Source: Kwok et al., 1990.

Table 7 Calculated Membrane Diffusivity and Size of Impermeable Gel Core with Carbonic Anhydrase as Diffusing Protein

PLL-alginate reaction time (mins)	Capsule diameter (μm)	Apparent diameter of impermeable gel core (μm)	Apparent volume fraction of intracapsular free space	Apparent membrane diffusivity ($\times 10^{12}$)(m²/min)
3	1,000	0	1,000	45.2
6	1,000	621	0.754	28.5
10	1,000	752	0.562	22.5
20	1,000	871	0.319	13.5

M_v of PLL = 21,000. M_v of carbonic anhydrase = 29,000. Capsule diameter arbitrarily set at 1,000 μm.
Source: Kwok et al., 1990.

in water, which range from 6.1 $\times$ 10^{-7} to 26.0 $\times$ 10^{-7} cm²/s. (Doherty and Benedek, 1974; Neal et al., 1984). Similar results were observed with carbonic anhydrase (Table 7). One possible explanation for our low values for D is that in our calculations, we assumed that the membrane thickness, $(R_2 - R_1)$, was 5 micrometers. This number was based on experimentally determined values reported earlier (Goosen et al.,

1985). Since $(R_2 - R_1)$ is a factor affecting membrane diffusivity, an error in the assumed membrane thickness would affect our calculated value for D. Another possible cause for these low values comes from the second assumption that the intracapsular region is well mixed. In fact, for this assumption to be valid, the diffusivity of albumin in the intracapsular region with a permeable core would have to be at least 100 times larger than that in the membrane (unpublished simulation results). If, however, a small permeable shell and a large impermeable gel core are present in the intracapsular region, the mass transport resistance in the liquid alginate phase will be greatly reduced, and the intracapsular region between the membrane and core will approach the "well-mixed" state as the core size becomes larger, even though the diffusivity in the intracapsular region may be only an order of magnitude larger in the membrane. The latter case should be more applicable to our system since an impermeable gel core was found in most cases. In any event, this resistance offered by the liquid alginate phase was lumped into the membrane resistance in our model and this would certainly bring down the calculated value of diffusivity of albumin in the membrane. This does not mean that the diffusivity of albumin in the membrane will increase with increasing core size, since the structure of the membrane changes with reaction time as well. The presence of this liquid alginate resistance simply means that the calculated apparent membrane diffusivities shown in Tables 6 and 7 should be lower than the real values, and that the larger the core size, the closer is the apparent value to the real value.

The apparent size (or diameter) of the impermeable gel core, on the other hand, increased as the duration of PLL-alginate reaction time increased. In the case of carbonic anhydrase ($MW = 29,000$) diffusing into the capsules, if we set the diameter of the microcapsule at 1,000 microns, (i.e., radius, r, in model equation is set at 500 microns), then the calculated size of the impermeable gel core increased from 0 to 871 microns as the PLL-alginate reaction time increased from 3 to 20 minutes (Figure 7). It is believed that the microcapsule membrane permeability also controls the extent, or ease, of core liquification during encapsulation (i.e., the rate of Ca^{2+}/Na^+ ion exchange reaction). The lower membrane diffusivity would have decreased the rate of sodium citrate diffusion into capsules and the rate of calcium citrate diffusion out of the capsules. We would thus expect the size of the impermeable gel core to increase with increasing alginate-PLL reaction time.

As the membrane diffusivity decreased, in the case of carbonic anhydrase diffusion studies, the size of the apparent impermeable gel core increased, but not to the same extent as observed with albumin. The

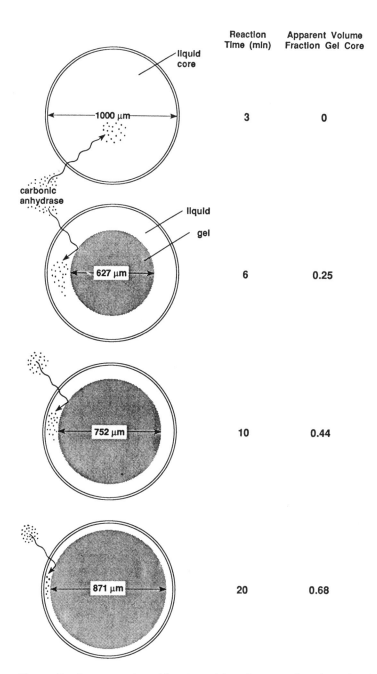

Figure 7 Apparent size of impermeable gel core as function of alginate-PLL reaction time with carbonic anhydrase as diffusing protein. (Adapted from Table 7).

calculated membrane diffusivities, with carbonic anhydrase as the diffusing substrate, were however up to eight times higher than the membrane diffusivities for the diffusion of albumin. In the case of microcapsules prepared with a PLL-alginate reaction time of 6 minutes, the membrane diffusivity increased from 6.7×10^{-12} m^2/min with albumin as the diffusing solute, to 28.5×10^{-12} m^2/min for carbonic anhydrase diffusion. This is not unexpected since carbonic anhydrase is a much smaller molecule (factor of 2) than albumin. An alternative explanation for the lower diffusivity of albumin is the fact that albumin in solution is negatively charged. Alginate (and perhaps the membrane) is also negatively charged. If the charge on albumin was greater than on carbonic anhydrase, then this would have resulted in a lower diffusivity for albumin.

An interesting observation from these studies was that the apparent gel core sizes appeared to be dependent on the size of the diffusing protein (Tables 6 and 7). Our results thus suggest that the pore size in the intracapsular gel phase may gradually decrease toward the center of the capsule (Figure 8). Comparing the equilibrium or final protein con-

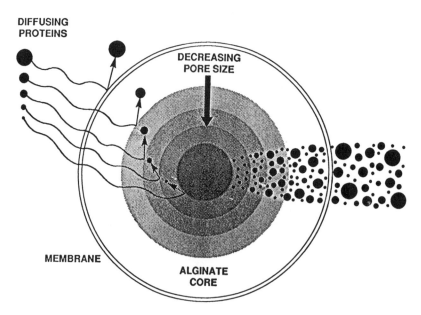

Figure 8 Proposed microcapsule model showing a partially impermeable gel core with decreasing pore size towards center of capsule. (Kwok et al., 1990)

centrations, higher extracapsular equilibrium absorbance readings were obtained with albumin than carbonic anhydrase. This indicates that fewer albumin molecules than carbonic anhydrase molecules would have diffused from the bulk solution into the capsules even though the total volumes of capsules and solution were the same in both cases. The intracapsular effective volume would therefore have been greater for the carbonic anhydrase than for albumin. According to a study by Tanaka et al., (1983), the diffusion of carbonic anhydrase into calcium alginate gel will depend on the pore size in the gel. Our experimental and theoretical evidence thus suggests that there may be an increase in the number and degree of crosslinks between alginate chains (caused by clacium ions) towards the centre of the capsule. The calcium ion concentration would therefore be expected to be the highest at the capsule center. The results also show that the effective volume is actually a function of the size of diffusing species with the gel pore size decreasing toward the center of the capsule.

Modeling Cell Growth

A dynamic two-dimensional model was also developed in our laboratory for cells cultured in stationary microcapsules (Yuet et al., 1992). The model generated results in close agreement with experimental data. Examination of the simulation results revealed a delicate balance between supply and demand of nutrients and oxygen. Low cell density was found in the central region of a microcapsule since the supply of oxygen and nutrients is reduced by the consumption of these nutrients by cells located near the capsule membrane (Figure 9). A high specific growth rate was found in the top layer of the cell population, where the supply of oxygen and nutrients is abundant. According to the simulation result, with microcapsules of 1020 μm in diameter and a capsule to medium volume ratio of 10%, only about 80% of the volume of a microcapsule is occupied by cells after 14 days of culture.

Using the model in place of actual experiments, the effects of operating variables such as capsule radius and membrane thickness on the intracapsular cell density was investigated (Yuet et al., 1992). For stationary microcapsules with low viscosity intracapsular liquid, it was found that capsule radius, capsule loading and medium change time had the most significant effects on the final intracapsular cell density at (14 days) and the fraction change in cell number per unit volume of culture medium. The diffusivity of small nutrients such as glucose and glutamine through the capsule membrane, however, did not seem to have any

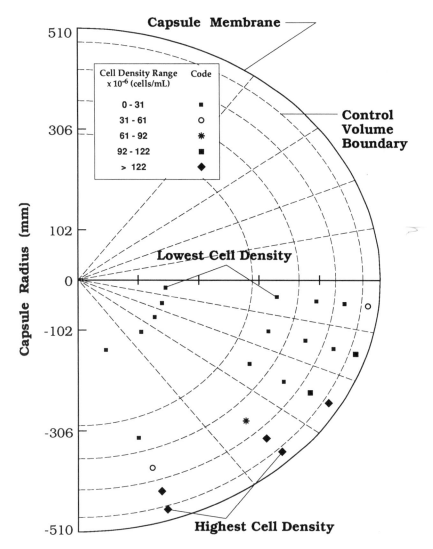

Figure 9 Simulated cell distribution in a microcapsule after 7 days of culture. Capsule diameter = 1020 μm. Highest cell density is found at the bottom of capsule near membrane.

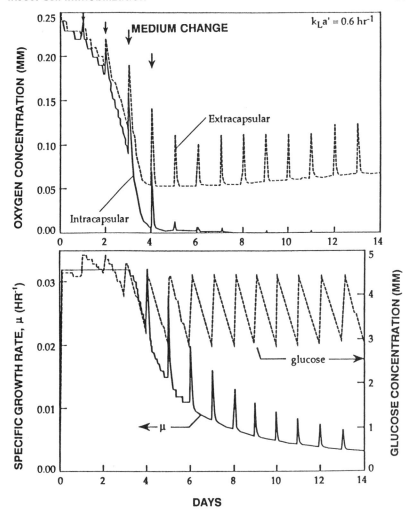

Figure 10 Simulated time variation of oxygen concentration in the medium and the interior phase and the specific growth rate and intracapsular glucose concentration for $K_L a' = 0.6$ hr^{-1}. Microcapsules with fluid intracapsular liquid in suspension. The medium was changed every 24 hours.

influence on these two response variables. The model was also adapted to simulate other scenarios of cell growth such as in gel beads and suspended microcapsules. The simulated time course of oxygen concentration and specific growth rate revealed a complicated interaction between material transport and cell growth kinetics (Figure 10). With the mass transfer coefficient for oxygen transfer $(K_L a')$ into the medium equal to 4.0 hr^{-1}, for instance, it was found that the specific growth rate of the microencapsulated cells was controlled, either separately or simultaneously, depending on the time in the culture period, by the supply of glucose and oxygen. When the value of $K_L a'$ was reduced to 0.6 hr^{-1} however, oxygen supply appeared to be the sole factor affecting the specific growth rate.

CONCLUDING REMARKS

Our biocompatibility and cell culture studies suggested that there is a mass transport limitation in the culture of insect cells immobilized in alginate capsules. This problem can apparently be bypassed by employing dilute ($<$ 1% w/v) sodium alginate solution in the immobilization procedure. It is unclear at this time what the rate limiting (nutrient) macromolecule is in the survival of immobilized insect cells.

A diffusion model was developed for the microcapsule system that successfully describes experimental diffusion profiles and which might be extended to other systems. Based on a comparison of calculated and experimental results, a physical microcapsule model is proposed for the alginate-PLL system, consisting of a semipermeable capsule membrane containing a partially impermeable alginate core with a decreasing pore size towards the centre of the capsule. This physical structure gives a fundamental insight into the diffusion restrictions for macromolecules. It was found, for example, that the ability of proteins to diffuse into the microcapsule was not only controlled by the membrane diffusivity, but also apparently by the amount of alginate remaining in the system. Our model can help determine the relative size of the impermeable core and thus should be able to predict cell viability in the microcapsules. The presence of a partially impermeable core might hinder the commercial use of microcapsules in cell culture engineering by decreasing productivity, and may affect clinical application of encapsulated cells/tissue in transplantation by affecting long-term (*in vivo*) cell viability. Furthermore, the size of the impermeable core might also change after *in vivo* transplantation. This would affect the ability of nutrients to reach the encapsulated cells.

What about commercial applications of cell immobilization technology? For animal cells, why culture immobilized cells if they can be grown just as well in suspension culture? If the decision is to use immobilized cell culture, should a membrane system be employed or a microporous bead system? The answers to these questions depend on the particular application and especially on the desired product. Compared to microcapsules, microporous beads are easier to use and to scale-up (i.e., packed and fluidized beds). Microcapsules and hollow fibres, on the other hand, provide for product concentration, although at a cost of increased complexity. Membrane technology is only one of many tools that can be used to produce and recover high value biologicals. In the ideal case it would be advantageous to grow cells and recover product at the same time. This may not always be possible, economically or techni-cally. The concerns often raised about the application of membrane technology to cell cultivation, (i.e., too complicated, too unreliable, and too costly) are in part valid. One of our aims should therefore be to simplify cell immobilization techniques. However, one cannot take new technology off the shelf and use it efficiently without understanding how it works. Technical staff, for instance, need to be trained in the proper use of new devices and techniques in cell cultivation. In conclusion, as our fundamental understanding of cell immobilization technology contin-ues to grow over the next decade, we can expect it to be used more effectively to solve problems in the biochemical and biomedical engineer-ing fields.

NOMENCLATURE

A_o	initial absorbance
A_t	absorbance at time t
A_1	inside membrane surface area, m^2
A_2	outside surface area of a microcapsule, m^2
A_2'	total outside surface area of microcapsules, m^2
C_b	protein concentration in the bulk solution, kg/m_3
C_i	concentration of protein in the alginate core, kg/m^3
C_m	concentration of protein in the membrane, kg/m^3
D_{ip}	diffusivity of protein in the alginate core, m^2/s
D_{mp}	diffusivity of protein in the membrane, m^2/s
F	effective volume fraction
M_b	amount of protein in the bulk solution, kg
M_i	amount of protein in the alginate core, kg
N	number of microcapsules
r	radial dimension, m
R_1	inner radius of a microcapsule, m
R_2	outer radius of a microcapsule, m
t	time, h
V_b	volume of the bulk solution, m^3

V'_c total volume of microcapsules, m^3
V_f effective or free volume between the membrane and the imperme-
          able fraction of the core in a microcapsule, m^3
V_i total intracapsular volume, m^3

ACKNOWLEDGMENTS

The financial assistance of the Natural Sciences and Engineering Research Council of Canada is gratefully acknowledged. The insect cell immobilization work and the mathematical modeling studies were performed by Mr G. King and Mr W. Kwok respectively. The insect cell cultivation project is a collaborative effort with Professor A. J. Daugulis, Professor P. Faulkner, and Professor T. Harris. The author would like to thank Professor B. Bellhouse of the University of Oxford for the use of office and laboratory space during his sabbatical, and Mrs W. Claye for typing the manuscript.

REFERENCES

Aebischer, P. (1990). Proceedings of ACS National Meeting, Boston, Massachusetts.

Allan, G. G., Altman, L. C., Bensinger, R. E., Ghosh, D. K., Hirabayashi, Y., Neogi A. N., and Neogi, S. (1984). Biomedical applications of chitin and chitosan, *Chitin, Chitosan and Related Enzymes,* (J. P. Zikakis, ed.), Academic Press, New York, pp.119–133.

Alper, J. (1984). The Stradivarius formula. Better music through chemistry, *Science, 84:*36–43.

Carslaw, H. S., and Jaeger, J. C. (1959). *Conduction of Heat in Solids,* Clarendon Press, Oxford.

Chang, T. M. S. (1964). Semipermeable microcapsules, *Science, 146:*524–525.

Chang, T. M. S., Macintosh, F. C., and Mason, S. G. (1966). Semipermeable aqueous microcapsules, *Can. J. Physiol. Pharmacol., 44:*115.

Chresand, T. J., Dale, B. E., Hanson, S. L., and Gillies, R. J. (1988). A stirred bath technique for diffusivity measurements in cell matrices, *Biotech. Bioeng., 32:*1029–1036.

Cook, W. H., and Smith, D. B. (1954). Molecular weight and hydrodynamic properties of sodium alginate, *Canadian J. Biochem. Physiology, 32:*227–239.

Crank, J. (1975). *The Mathematics of Diffusion,* 2nd Ed., Clarendon Press, Oxford.

Doherty, P. and Benedek, G. B. (1974). The effect of electric charge on the diffusion of macromolecules, *J. Chem. Phys., 61(12):*5426–5434.

Flynn, G. L., Yalkowsky, S. H., and Roseman, T. J. (1974). Mass transport phenomena and models: Theoretical concepts, *J. Pharm. Sci., 65(4):*479–510.

Gharapetian, H., Maleki, M., Davies, N. A., and Sun, A. M. (1986). Polyacrylate membranes for encapsulation of viable cells, *Polym. Mater. Sci. Eng., 54:*114–118.

Goosen, M. F. A., O'Shea, G. M., Gharapetian, H. M., Chou, S., and Sun, A. M. (1985). Optimization of microencapsulation parameters: Semipermeable microcapsules as a bioartificial pancreas, *Biotechnology and Bioengineering, 27:*146–150.

Goosen, M. F. A., King, G. A., Daugulis, A. J., and Faulkner, P. (1990). Multiple Membrane Encapsulation. U.S. Patent No. 4,942,129.

Green, B. K. (1955). Pressure-Sensitive Record Materials. U.S. Patent No. 2,712,507.

Guiseley, K. B. (1989). Chemical and physical properties of algal polysaccharides used for cell immobilization, *Enzyme Microb. Technol., 11:*706–716.

Guo, Y., Lou, F., Peng, Z., and Korus, R. A. (1990). Kinetics of growth and a-amylase production of immobilized *Bacillus subtilis* in an airlift bioreactor, *Biotech. Bioeng., 35:*99–102.

Hackel, V., Klein, J., Megret, R., and Wagner, F. (1975). Immobilization of microbial cells in polymeric matrices, *Eur. J. Appl. Microbiol., 1:*291.

Heath, C., and Belfort, G. (1987). Immobilization of suspended mammalian cells: Analysis of hollow fibre and microcapsule bioreactor, *Adv. Biochem. Eng/Biotech, 34(1):*31.

Hirano, S. (1978). A facile method for the preparation of novel membranes from n-acyl and n-arylidene–chitosan gels, *Agric. Biol. Chem., 42(10):*1939–1940.

Itamunoala, G. F. (1987). Effective diffusion coefficients in calcium alginate gel, *Biotech. Progress, 3 (2):*115–120.

Iwata, S., Narui, T., Takahasih, K., and Shibata, S. (1985). Preparation of o-(carboxymethyl) cellulose (cmc) of high degree of substitution, *Carbohydrate Research, 145:*160–162.

Kierstan, M., and Bucke, C. (1977). The immobilization of microbial cells, subcellular organelles, and enzymes in calcium alginate gels. *Biotechnol. Bioeng., 9:*387.

King, G. A., Daugulis, A. J., Faulkner, P., and Goosen, M. F. A. (1987). Alginate-polylysine microcapsules of controlled membrane molecular weight cut off for mammalian cell culture engineering, *Biotech. Progress, 3(4):*231–240.

King, G. A., Daugulis, A. J., Goosen, M. F. A., Faulkner, P., and Bayly, D. (1989). Alginate concentration: A key factor in growth of temperature-sensitive baculovirus-infected insect cells in microcapsules, *Biotech. Bioeng., 34(8):*1085–1091.

Knorr, D., Beaumont, M. D., and Pandya, Y. (1987). Polysaccharide copolymers for the immobilization of cultured plant cells, *Biotechnol. Food Ind., Proc. Int. Symp.,* 389–400.

Kwok, W. Y., Kiparissides, C., Yuet, P., Harries, T. J., and Goosen, M. F. A. (1990). Mathematical modelling of protein diffusion in microcapsules; A comparison with experimental results, *Canadian J. Chem. Eng, 68.*

Lakshminarayanaiah, N. (1969). *Transport Phenomena in Membranes,* Academic Press, New York, pp. 4–6.

Lapasin, R., Zanetti, F., and Paoletti, D. 1988. *Suppl. Rheologia Acta., 27:*422–424.

Lim, F., and Sun, A. M. (1980). Microencapsulated islets as bioartificial endocrine pancreas, *Science, 210:*908–910.

Lim, F. (1982). Encapsulation Biological Materials. U.S. Patent No. 4,352,883.

Lim, F. (1984). Microencapsulation of living cells and tissues—Theory and practice, in *Biomedical Applications of Microencapsulation,* pp.137–154.

Martinsen, A., Skjak-Braek, G., and Smidsrod, O. (1989). Alginate as immobilization material: 1. Correlation between chemical and physical properties of alginate gel beads, *Biotech. Bioeng., 33:*79–89.

McCarthy, A., McHale, A. P., and McHale. M. L. (1988). Studies on the encapsulation of rat pancreatic cell line AR 4SJ, *Biochem. Transactions, 17:*393–394.

McKnight, C. A., Penny, C., Ku, A., Sun, D., and Goosen, M. F. A. (1988). Synthesis of Chitosan-Alginate Microcapsule Membranes, *Journal of Bioactive and Compatible Polymers., 3:*334–355.

Morgensen, A. O., and Vieth, W. R. (1973). Mass transfer and biochemical reaction with semipermeable microcapsules. *Biotechnol. Bioeng., 15:*467–481.

Mosbach, K. (1984). New immobilization techniques and examples of their applications, in *Annals of the New York Academy of Sciences Vol. 434: Enzyme Engineering 7,* (A. I. Laskin, G. T. Tsao, and L. B. Wingard Jr., eds.), New York Academy of Sciences, New York, 239.

Muzzarelli, R. A. A. (1977). *Chitin,* Pergamon of Canada Ltd., Toronto.

Neal, D. G., Purich, D., and Cannell, D. S. (1984). Osmotic susceptibility and diffusion coefficient of charged bovine serum albumin, *J. Chem. Phys., 80* (7), 3469–3477.

Percival, E., and McDowell, R. H. (1967). *Chemistry and Enzymology of Marine Algal Polysaccharides,* Academic Press, New York.

Posillico, E. G. (1986). Microencapsulation technology for large-scale antibody production, *Bio/Technology, 4*(2) 114–117.

Reilly, A., Antognetti, G., Wesolowski, Jr. J. S., and Sakorafas. (1990). The use of microcapsules for high density growth of cells for infiltrating lymphocytes and other immune reactive T cells. *Immunological Methods, 126,* 273–279.

Rha, C. K. (1985). Process for Encapsulation of Active Material. European Patent Application No. 152,898.

Rha, C. K., Rodrigues-Sanches, D. (1988). Encapsulated Active Material System. U.S. Patent No. 4,749,620.

Rupp, R. G. (1985). *Large-Scale Mammalian Cell Culture,* (J. Feder and W. R. Tolbert, eds.), Academic Press, New York.

Sefton, M. V., and Broughton, R. L. (1982). Microencapsulation of erythrocytes, *Biochimica et Biophysica Acta, 717:*473–477.

Sefton, M. V., Dawson, R. M., Broughton, R. L., Blysniuk, J., and Sugamori,

M. E. (1987). Microencapsulation of mammalian cells in a water insoluble polyacrylate by coextrusion and interfacial precipitation. *Biotechnology and Bioengineering, 29*:1135–1143.

Sefton, M. V., Crooks, C. A., Horvath, V., and Kharlip, L. (1990). HEMA–MMA Copolymer for the Microencapsulation of Mammalian Cells, Research Report, University of Toronto.

Shahar, A., and Reuveny, S. 1987. Nerve and muscle cells on microcarriers in culture, *Advances in Biochemical Engineering/Biotechnology, Vol. 34*, (Fiechter, ed.), Springer-Verlag, pp.33–35.

Shiotani, T., Shiiki, Y., Kimura, T., Kako, M., and Hayashi, Y. (1986). Encapsulation of Unstable Cells and Substances with Polyanionic Polysaccharides and Chitin Derivatives. Japanese Patent No. 61, 11, 139.

Shukla, R., Kang, W., and Sirkar, K. K. (1989). Acetone-butanol-ethanol production in a novel hollow fiber fermentor-extractor, *Biotech. Bioeng., 34*, 1158–1166.

Skjak-Braek, G. (1988). *Biosynthesis and Structure–Function Relationships in Alginates,* Division of Biotechnology, University of Trondheim NTH, Norway.

Smith, N. A., Goosen, M. F. A., King, G. A., Faulkner, P., and Daugulis, A. J. (1989). Toxicity analysis of encapsulation solutions and polymers in the cultivation of insect cells, *Biotech. Letters, 3* (*1*):61–66.

Strauss, S. (1988). Bug harvest on way, chemist predicts, *The Globe and Mail*, p.A3, 7 June.

Sun, A. M., and O'Shea, G. M. (1985). Microencapsulation of living cells—A long-term delivery system. *J. Control. Rel.*, 137–141.

Tanaka, H, Matsumura, M., and Valiky J. A. (1983). Diffusion characteristics of substrates in Ca-alginate gel beads, *Biotech Bioeng, 26*:52, 58.

Tanaka, H., Matsumura, M., and Veliky, I. A. (1984). Diffusion characteristics of substrates in Ca-alginate gel beads. *Biotech. Bioeng., 26*:53–58.

Tice, T. R., and Meyers, W. E. (1985). Encapsulated Cells, Their Method of Preparation and Use. European Patent Application No. 0,129,619.

Yalpani, M. (1988). *Polysaccharides,* Elsevier Science Publishing Company, Inc., Amsterdam.

Yoshioka, T., Hirano, R., Shioya, T., and Kako, M. (1990). Encapsulation of mammalian cell with chitosan-cmc capsule. *Biotech. Bioeng., 35*, 66–72.

Yuet, P. K., Goosen, M. F. A., and Harris, T. J. (1992). Mathematical modelling of animal cell growth in microcapsules. (Submitted)

5

Applications of Insect Cell Gene Expression in Pharmaceutical Research

Darrell R. Thomsen   *The Upjohn Company, Kalamazoo, Michigan*

Annette L. Meyer and Leonard E. Post   *Parke-Davis, Ann Arbor, Michigan*

5.1 INTRODUCTION

When investigating the therapeutic potential of a protein, it is necessary to have a sufficient quantity of the molecule available to study its biological activity. Similarly, discovery of chemical compounds as drugs often requires quantities of an enzyme or a receptor on which the drug is intended to act. Prior to the advent of biotechnology, usable quantities of proteins for analysis as therapeutic agents or therapeutic targets were available only by extraction from tissue, blood, or some other biological source. This often involved large amounts of tissue. In some cases when the protein of interest was present in the tissue at low abundance, it was impossible to purify useful amounts. The limiting step in the analysis of any protein has often been the availability of a protein source. Recombinant DNA technology has, in many cases, provided a solution to this problem. Once a gene is isolated it can be inserted into one of the many expression systems available. *Escherichia coli*, yeast, mammalian cells, or the baculovirus expression system (BEVS), and large quantities of the protein can be produced for purification and analysis.

This chapter will deal with one of these expression systems, the baculovirus expression system, and how we and others have used it to produce and analyze proteins as pharmacologic agents, vaccines, or therapeutic targets. Our work involves the expression of proteins and glycoproteins which are normally produced by mammalian cells. The baculovirus expression system has been shown to perform most of the posttranslational modifications required to make a biologically active protein, regardless of the source of the gene (Luckow, 1990; Luckow and Summers, 1988a; O'Reilly et al., 1992; Summers, 1988; Summers, 1989). Genes from plants, fungi, viruses, bacteria, and animals have been expressed and found to be active when produced in infected insect cells. The baculovirus expression system has provided us with a rapid method to produce substantial quantities of active proteins and glyco-proteins. It has been especially helpful in production of glycoproteins that we have been unable to make in other eucaryotic or procaryotic expression systems. To date we have inserted over forty different genes into the baculovirus AcNPV-E2. In every case the product of interest was expressed and appeared to be biologically active. The expression of five types of glycoproteins will be discussed: the first type involves the production of glycoproteins as vaccine antigens. Two antigens will be discussed, gp50 glycoprotein from pseudorabies virus (PRV) and an FG fusion glycoprotein from human respiratory syncytial virus (RSV). The other types of glycoproteins include: a therapeutic, tissue plasminogen activator (TPA); a hormone, nerve growth factor (NGF); and two poten-tial therapeutic targets, one an enzyme, renin, and the other a mem-brane receptor for interleukin-1 (IL–1R).

5.2 GENERAL EXPERIMENTAL PROCEDURES

The baculovirus expression system we use was obtained from M. D. Summers, and all manipulations of viruses and Sf-9 cells were performed using methods published by Summers and Smith (Smith et al., 1983; Summers and Smith, 1987). The biology of baculoviruses (Doerfler and Bohm, 1986; Faulkner, 1981; Granados and Federici, 1986a; Summers, 1978; Tanada and Hess, 1984; Blissard and Rohrmann, 1990, Granados and Federici, 1986b) and the use and applications of baculovirus expres-sion vectors (Luckow, 1990; Luckow and Summers, 1988a; O'Reilly et al., 1992; Summers, 1988; Summers, 1989; Miller, 1981; Miller, 1987; Miller et al., 1983; Miller et al., 1986; Kang, 1988; Cameron et al., 1989; Frazer, 1989; Maeda, 1989; Miller, 1988; Miller, 1989) have been re-viewed extensively by others.

Three different plasmids were used as expression vectors, pAC373 (Smith et al., 1985), pVL941 (Luckow and Summers, 1989), and pACYM1 (Matsurra et al., 1987). In all cases the gene to be expressed was cloned in at the BamHI site of the vector. The renin gene and gp50 gene were inserted into pAC373, the interleukin-1 receptor and nerve growth factor genes into pVL941, and the FG fusion gene into all three vectors. NGF, FG, and the IL-1 receptor were expressed by Sf-9 cells growing in serum-free medium (Inlow et al., 1989). Unless otherwise stated, Sf-9 cells growing in serum-containing TNM-FH (Hink, 1970), or serum-free medium, were infected with recombinant viruses at a multiplicity of infection (MOI) of 5 pfu per cell. Scale-up of Sf-9 cells for protein production was performed in spinner flasks (in serum-containing medium), in shake flasks, or in a 6 liter airlift fermentor (LH Fermentation, Hayward, CA) in serum-free medium. Optimal harvest times were determined for each protein. Specific parameters unique to the individual glycoproteins will be discussed in the pertinent sections.

Labeling experiments with either ^{14}C-glucosamine or ^{35}S-methionine were performed from 24 to 48 hours post infection. Although this is usually not the period of maximal protein production (optimal times are between 48 and 64 hours post infection), we felt that this interval would provide a low background of host protein synthesis as well as a sufficient amount of recombinant protein synthesis for detection. In addition, during this period production levels should be low enough that the posttranslational modifying systems would not be overwhelmed with viral or recombinant protein; and that the cells' processing machinery would not be damaged by the viral infection. Therefore the results should show an accurate picture of the processed protein.

5.3 PRODUCTION OF ANTIGENS AS SUBUNIT VACCINES

Animal Vaccines: Pseudorabies Virus

Pseudorabies virus (PRV), also know as Aujesky's disease virus, is a significant pathogen of swine (Gustafson, 1981), causing economic losses in the millions of dollars worldwide. Within an infected herd, mortality can be as high as 100% in suckling pigs, while adult animals may show only a mild or inapparent infection. Pregnant animals may abort or young may be stillborn. Fertility within an infected herd is normally reduced following the outbreak. As with other herpesvirus infections, animals that survive the initial infection carry the virus for life in a latent state. Under some conditions this virus can reactivate and infect unprotected animals.

There are several live or inactivated PRV vaccines currently used to control PRV infections. Even though some of these vaccines work exceptionally well (providing 100% protection against clinical symptoms), subunit vaccines are considered by some to be safer, since they do not involve the release of live virus into the environment. In addition, subunit vaccine production need not involve the antigenic alterations that can occur when viruses are inactivated for killed vaccines. The BEV system provided a rapid method to test the PRV glycoprotein gp50 as a subunit vaccine for pseudorabies.

The gp50 glycoprotein is an ideal candidate for a subunit vaccine. Antibodies to it neutralize the virus. Since it shares sequence homology with herpes simplex virus (HSV) gD (Petrovskis et al., 1986), and gD is known to be essential for virus replication (Johnson and Ligas, 1988), the virus should not be able to circumvent the immune response by deleting the gene. The gp50 glycoprotein (Ben-Porat and Kaplan, 1970) is a structural component of the virus envelope and is thought to be involved with viral attachment to host cell receptors. Many times surface glycoproteins are excellent targets for neutralization.

The gp50 gene was first described by Wathen and Wathen (1984). The gene has been sequenced (Petrovskis et al., 1986) and found to be a typical class 1 glycoprotein, having a hydrophobic signal and transmembrane domain and a hydrophilic cytoplasmic domain. It contains no *N*-linked glycosylation sites even though it can be labeled with glucosamine. In PRV infected vero cells it has a molecular weight of 60 kD, while in MVPK cells it has a molecular weight of only 50 kD. This is presumably due to decreased glycosylation in the latter cell type.

Our goal was to produce enough gp50 to test as a subunit vaccine. In the process of doing this we discovered some unique features about the ability of insect cells to add *O*-linked oligosaccharides to proteins. Although the results discussed concerning gp50 have been developed further by actual analysis of the oligosaccharide structures on gp50 (Thomsen et al., 1990), the data presented show a common approach that can be used for the analysis of a protein expressed in baculovirus.

Two forms of the gp50 gene were inserted into the transfer vector pAC373. One form codes for the complete gene and should remain within the infected cell. A second form, designed to be secreted, is a truncated version, with codons for the carboxyl terminal hydrophobic transmembrane and cytoplasmic domains deleted. The truncated form was engineered to simplify purification of the glycoprotein, should it be an effective vaccine. Both genes were inserted into the baculovirus AcNPV-E2. Two methods were employed to analyze the protein produced. The first tested the ability of the virus to produce gp50, while the second provided insight into processing of the glycoprotein.

A typical labeling experiment is shown in Figure 1. Sf-9 cells were infected with AcNPV-TPA (control), AcNPV-gp50 (expressing the intact protein), or AcNPV-gp50T (expressing the truncated protein). The cells were labeled with either ^{35}S-methionine or ^{14}C-glucosamine, followed by immunoprecipitation of the labeled proteins using a monoclonal antibody to gp50. Two forms of gp50 are produced, having molecular weights of 47 kD and 46 kD. Both forms of the glycoprotein incorporate a glucosamine label and remain within the cell, as expected. In the case of gp50T, 44 kD and 45 kD glycosylated proteins are present in infected cells with higher amounts present in the medium. A 43 kD nonglycosylated species is also present, but only in the cells. These results suggest that gp50 produced in insect cells is being posttranslationally modified. To analyze this further, we performed a pulse-chase experiment. This was expected to show an initial precursor that increases in molecular weight during the chase, much like that seen in PRV infected vero cells (Petrovskis et al., 1986). The results of this experiment are shown in Figure 2. Sf-9 cells were infected with AcNPV-gp50T. At 24 hours post infection the cells were pulsed for 15 minutes with ^{35}S-methionine followed by chase periods of the times shown. Labeled proteins were immunoprecipitated. At the end of the pulse period a 41 kD polypeptide is found in the cells. With additional time, increasing amounts of 43 kD and 44 kD species appear in the cell as the amount of the 41 kD form decreases. These two forms eventually chase into the medium. These results show that insect cells are indeed capable of posttranslationally modifying gp50, which has no sites for N-linked glycosylation. Later work (Thomsen et al., 1990) shows that this modification was addition of O-linked oligosaccharides to the protein backbone. The major O-linked structure found on gp50 made in insect cells is the monosaccharide GalNAC with lesser amounts of Galβ1-3GalNAC. No sialic acid was found. When gp50 is produced in PRV infected vero cells, it contains only the disaccharide Galβ1-3GalNAC, either substituted or unsubstituted with one or two sialic acid residues.

This difference in glycosylation by insect cells could have changed the antigenic structure of gp50 compared to that made in mammalian cells. To determine whether this was the case, gp50T produced by baculovirus and by a stably transformed Chinese hamster ovary (CHO) cell line (CHO gp50T) (A. L. Meyer and E. A. Petrovkis, unpublished results) were compared in a competition ELISA assay (Figure 3). This assay uses an antiserum produced by vaccinating a rabbit with a recombinant vaccinia virus expressing gp50 (Marchioli et al., 1987). The AcNPV-gp50T was shown to compete as effectively for binding to the polyclonal antibody as

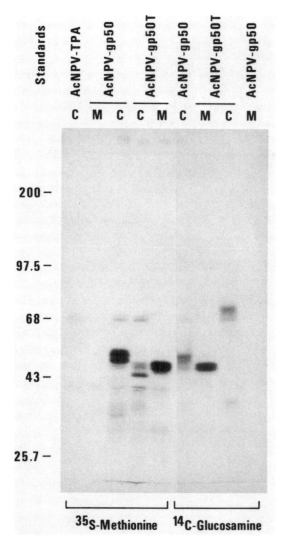

Figure 1 Synthesis of gp50 by recombinant baculoviruses. Sf-9 cells were infected with either AcNPV-gp50 or AcNPV-gp50T at a MOI of 5 pfu/cell. At 24 hr post infection, the infected cells were labeled with either ^{35}S-methionine (50 μC/ml) or ^{14}C-glucosamine (50 μC/ml). The samples were harvested after 24 hr labeling. The samples were immunoprecipitated with anti-gp50 monoclonal antibody 3A-4 and loaded onto a 12.5% SDS polyacrylamide electrophoresis gel. An autoradiogram is shown. Lanes labeled C are immunoprecipitations of extracts from infected cells, and lanes labeled M are immunoprecipitations of media from infected cells.

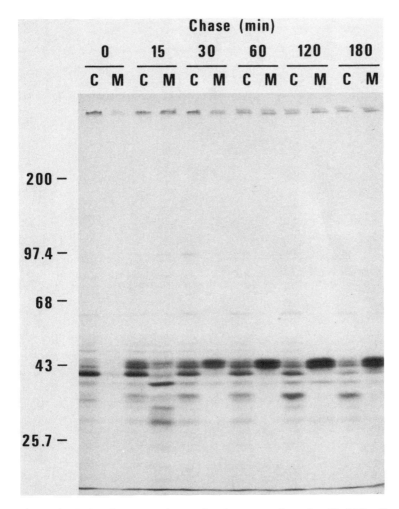

Figure 2 Pulse-chase experiment showing processing of gp50. Sf-9 cells were infected at a MOI of 5 pfu/cell with AcNPV-gp50T. At 24 hr post infection, cells were washed two times with Grace's medium minus methionine, then pulse-labeled for 15 minutes with 100 μC/ml ^{35}S-methionine in Grace's medium minus methionine. The label was removed and chased using standard Grace's medium. At indicated times after beginning of the chase, cells (C) and medium (M) were harvested. The samples were immunoprecipitated with anti-gp50 monoclonal antibody 3A-4 and loaded onto a 12.5% SDS polyacrylamide electrophoresis gel. An autoradiogram is shown.

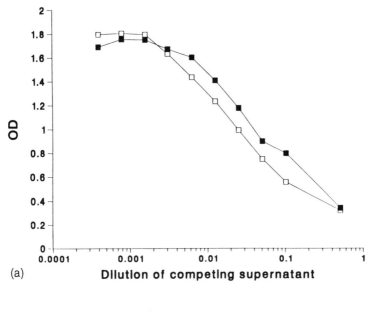

(a)

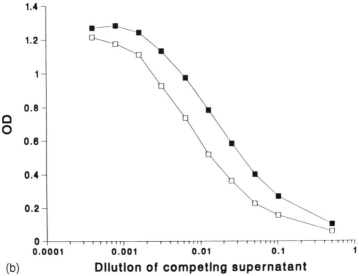

(b)

Figure 3 Competition ELISA between AcNPV-gp50T and CHO gp50T. The primary antigen, diluted 1/10 in 50 mM carbonate buffer, pH 9.6, was adsorbed to the wells of a polystyrene microfilter plate. The competing antigen was serially diluted twofold and added to the coated wells at the same time as diluted

Table 1 Protection of Swine from PRV Challenge by Immunization with AcNPV-gp50 Infected Cells

Pig	Treatment	Days virus isolation[a]	Clinical symptoms[b]
1	gp50[c]	4	D(9)[d]
2	gp50	4	1[e]
3	gp50	5	1
4	gp50	2	1
5	None	6	D(8)
6	None	6	D(8)
7	None	9	1
8	None	3	D(6)

[a]Virus isolation from nasal swabs.
[b]Clinical signs included anorexia, incoordination, inability to rise, and convulsions.
[c]Vaccinated once with 2×10^7 AcNPV-gp50.
[d]Death (day post challenge).
[e]Days.

CHO gp50T, indicating that the same epitopes are present on the two proteins. The only difference appears to be in the amounts of gp50T produced, with the insect cells producing more.

The gp50 produced by baculovirus was tested in swine for its ability to protect against a lethal PRV challenge (Table 1). Six week old pigs were given a single dose of 2×10^7 AcNPV-gp50 infected cells in complete Freund's adjuvant. Twenty-one days later the animals were challenged intranasally with a lethal dose of PRV and observed for 10 days. Insect cell produced gp50 was capable of protecting swine from death, although protection was not as complete as that provided by live vaccines. All the pigs immunized with gp50 showed signs of clinical disease. It may be possible to increase the immune response by the use of booster

---

rabbit vaccinia gp50 serum and incubated for 2 hr. After washing the plate, horseradish peroxidase-conjugated goat anti-rabbit IgG was incubated on the plate for 1 hr. An additional wash step was followed by incubation with a substrate, H_2O_2, and chromogen, *o*-phenylene diamine. The reactions were stopped with 1 M H_2SO_4 and absorbance read at 492 nm. (a) Primary antigen was AcNPV-gp50T; competing antigen was either CHO gp50T (□) or AcNPV-gp50T (■). (b) Primary antigen was CHO gp50T; competing antigen was either CHO gp50T (□) or AcNPV-gp50T (■).

inoculations, larger doses of gp50, or purification of the gp50 prior to immunization. Alternatively, several glycoproteins from PRV, such as gp50 and gIII, could be combined in a multivalent subunit vaccine. This approach may be feasible and should be tested. Unfortunately, cost becomes a problem with these options. At present this cost factor and unfavorable efficacy compared to live vaccines have halted any more experimentation on gp50 as a subunit vaccine.

Other herpesvirus surface glycoproteins have been expressed in baculovirus infected insect cells: glycoprotein D of herpes simplex virus (HSV) (Krishna et al., 1989), which is homologous to PRV gp50, and gB of human cytomegalovirus (Wells and Compans, 1990a). Like gp50, these were less extensively glycosylated than their mammalian-derived counterparts. The cleavage of glycoprotein B (150 kD) to its normal membrane bound form, a 116 kD-55 kD disulfide linked complex, apparently occurs only at reduced levels. This reduced glycosylation or partial cleavage does not appear to affect the immunogenic properties of these glycoproteins. Both are capable of eliciting a neutralizing antibody response in animals, and HSV glycoprotein D provides protection from virus challenge when used as a subunit vaccine.

A Human Vaccine: Respiratory Syncytial Virus

Human respiratory syncytial virus (RSV), a member of the paramyxoviridae family, is the predominant infectious agent of acute lower respiratory tract infections in infants and young adults (Kingsbury et al., 1978). The RSV genome encodes two glycoproteins, F and G (Dubovi, 1982; Huang et al., 1985; Peeples and Levine, 1979; Walsh and Hruska, 1983). The F glycoprotein is made as a 68 kD precursor, FO, which is proteolytically cleaved into disulfide linked subunits, F1 at 48 kD and F2 at 20 kD. This cleavage is required to activate the fusion activity associated with the F glycoprotein (Gruber and Levine, 1983). When the F glycoprotein is expressed in insect cells it undergoes only partial cleavage at this site. In addition, a secondary cleavage site is recognized by insect cells resulting in the production of two forms of F1, F1a and F1b (Wathen et al., 1989a). The G glycoprotein produced in mammalian cells is associated with viral attachment (Levine et al., 1987). It is heavily glycosylated, with both *N*-linked and *O*-linked oligosaccharides (Lambert, 1988; Wertz et al., 1985). This glycoprotein is made as a 33 kD precursor which is processed to a 90 kD form. The majority of this increase is reported to be due to the addition of *O*-linked oligosaccharides. Expression of the G glycoprotein in baculovirus has not been reported.

The F and G glycoproteins are the principal targets of the host's neutralizing antibody response to RSV infection (Levine et al., 1988). Antibodies to F prevent cell fusion and spread of the virus from cell to cell, while antibodies to G prevent attachment of the virus to host cell receptors. Since antibodies to both proteins are important in controlling infection, a subunit vaccine should contain both glycoproteins. Rather than producing each glycoprotein separately, we decided to fuse them to make a chimeric FG glycoprotein which would, we hoped, have the immunogenic properties of both, yet be easier and more cost effective to produce (Wathen et al., 1989b).

The engineered FG gene encodes the first 489 amino acids of the F glycoprotein, including the signal sequences but lacking the carboxyl terminal anchor region. The F region was fused to the G glycoprotein gene at codon 97 of G and continues to codon 279. The G portion of the chimeric gene does not encode either the amino terminal signal or the anchor sequences. Since the anchor sequences of both F and G are absent in the chimera, the product should be secreted into the medium of infected cells. The F region includes most of the protein backbone which is exposed at the surface of the virion and is a likely immunologic target. The G portion includes most of the major neutralizing epitopes recognized by infected animals (Olmsted et al., 1989).

Attempts to express the FG fusion glycoprotein in mammalian cells using stably transformed CHO cells or a vaccinia virus vector met only limited success. Only the vaccinia virus expression system produced FG at a level comparable to that of baculovirus. Rather then dealing with a live vaccinia virus vector, the subunit approach was taken, and we opted for production of FG using the BEV system.

Fusing two glycoproteins could lead to many problems at the expression, processing, and structural levels. Structurally, many of the epitopes on F and G could have been destroyed in the fusion process due to removal by design or by improper folding. These potential structural alterations could also interfere with secretion of FG from infected cells.

FG was initially cloned into pAC373. This vector produced about 500 μg of purified FG per liter of infected cells (2×10^6 cells/ml). FG was also cloned into pACYM1 and pVL941. These vectors have been reported to produce higher levels of some proteins than pAC373. In our hands FG production remained the same regardless of the vector employed.

The processing of FG (Wathen et al., 1989b) is similar to that of F produced in insect cells (Wathen et al., 1989a). The F portion of the chimera is cleaved to form F1G and F2 subunits. Since the FG glycoprotein migrates as a diffuse band on SDS-PAGE gels, the possibility still exists that the F portion of the molecule also undergoes secondary cleavage yielding F1aG and F1bG products. This possibility has not yet been tested.

Several viral surface glycoproteins that are normally cleaved during expression have been found to remain uncleaved or to undergo incomplete cleavage in insect cells. No proteolytic cleavage was seen in insect cell produced JHM gpE2 (Prehaud et al., 1989), influenza A HA (Possee, 1986), or parainfluenza type 3 F (Ray et al., 1989). Inefficient cleavage was found to occur in rabies F (Prehaud et al., 1989), cytomegalovirus gB (Wells et al., 1990b), and fowl plague HA (Kuroda et al., 1986). It appears that insect cells have the necessary factors to cleave many of these proteins, however these factors are not efficient enough to complete processing.

To test for structural as well as antigenic integrity of the FG glycoprotein, a comparison was made between FG produced in insect cells and F and G produced in RSV infected mammalian cells using a battery of monoclonal antibodies against the F and G glycoproteins (Wathen et al., 1989b). Reactivity with these monoclonal antibodies was analyzed by ELISA and immunoprecipitation. Of the monoclonal antibodies tested, all but one reacted with the insect cell FG in the ELISA. This antibody, as well as one other, also failed to immunoprecipitate FG from infected insect cells. This shows that the FG glycoprotein produced in insect cells retains most of the epitopes present on the authentic F and G molecules. The two antibodies that failed to react also reacted poorly with RSV produced F and G. These data were promising and warranted beginning animal experiments.

Antigenic integrity is extremely important. In the 1960s a formalin inactivated RSV vaccine was administered to children (Kim et al., 1969). This vaccine failed to protect and in some cases led to exacerbation of disease in children exposed to RSV. The reasons for this are unknown. One theory is that formalin denatured one of the virion proteins, destroying an essential neutralizing epitope or epitopes (Murphy et al., 1986). Before FG, or any other subunit, can be used as a vaccine in human trails, it is important to determine if this same exacerbation of the disease will occur.

Efficacy of FG was first tested in cotton rats (Wathen et al., 1989a; Brideau et al., 1989). FG was purified using monoclonal antibody columns and formulated in alum. Following immunization, animals were challenged with virulent RSV. FG was found to function well as a subunit vaccine, in a typical dose response matter. Protection could be achieved with a very low dose; 50 μg of purified FG provided 100%

protection even though the neutralization titer was low. At 10 μg there was a significant reduction in the replication of the virus in the lungs. These data were encouraging, and the FG glycoprotein produced in baculovirus infected insect cells is now a candidate for an RSV vaccine in children and is in preclinical development.

Table 2 shows a list of viral surface glycoproteins that have been expressed using the baculovirus expression system. In all cases these glycoproteins were found to be less glycosylated, as measured by molecular weight relative to the same protein made in mammalian cells. This did not appear to affect the ability of the glycoproteins to raise neutralizing antibody or a protectice immune response when tested. Full length glycoproteins were found on the surface of the cell (except for Dengue virus glycoprotein E), showing that transport as well as insertion of the glycoprotein into the membrane is occurring in insect cells. Only three of the glycoproteins, RSV FG and F, and HIV gp160, were purified prior to being tested for their ability to protect animals from challenge virus. In most cases, the immunogen was a crude mixture of infected cell proteins, a small fraction of which was the recombinant protein. Even under these less than optimal conditions many of the glycoproteins were found to afford some protection.

5.4 EXPRESSION OF AN ENZYMATICALLY ACTIVE GLYCOPROTEIN

Tissue plasminogen activator (TPA) is a member of a class of serine proteases that cleaves the proenzyme plasminogen into plasmin. Plasmin then degrades the fibrin network of blood clots (Christman et al., 1977; Collen, 1980). TPA has been developed and licensed by the Food and Drug Administration as a thrombolytic agent (Collen et al., 1984). TPA is a glycoprotein that is synthesized as a single polypeptide and cleaved during fibrinolysis into a two chain disulfide linked form (Rijken and Collen, 1981; Wallen et al., 1983). It has been expressed in several different systems: *E. coli* (Pennica et al., 1983), *Saccharomyces cerevisiae* (MacKay, 1988; LeMontt et al., 1985), mammalian cells (Sambrook et al., 1986; Kaufman et al., 1985; Collen et al., 1984; Browne et al., 1985), and baculovirus infected insect cells (Jarvis and Summers, 1989; Luckow and Summers, 1988b; Furlong et al., 1988; Steiner et al., 1988). Mammalian cells are capable of producing exceptionally high levels of active TPA. CHO cells transfected with the TPA cDNA and amplified with methotrexate can produce around 8 μg of TPA/10^6 cells/day for up to 5 weeks in continuous culture (D. P. Palermo and H. D. Fischer,

Table 2 Expression of Viral Surface Glycoproteins Using the Baculovirus Expression System

Virus	Gene	Cellular location	Neutralization[a]	Protection[b]	References
Parainfluenza-3	HN	Cell surface	+	Cotton rats	Van Wyke-Coelingh et al., 1987
Parainfluenza-3	F	Cell surface	+	Hamsters	Ray et al., 1989
Respiratory syncytial	F	N.D.	+	N.D.	Wathen et al., 1989a
Respiratory syncytial	F(truncated)	Secreted	+	Cotton rats	Wathen et al., 1989a
Respiratory syncytial	FG(chimeric)	Secreted	+	Cotton rats	Wathen et al., 1989b
Fish rhabdovirus	IHNVgp	N.D.	N.D.	N.D.	Koener and Leong, 1990
Vesicular stomatitis	G	Cell surface	N.D.	N.D.	Bailey et al., 1989
Rabies	G	Cell surface	+	Mice	Prehaud et al., 1989
Bovine coronavirus	HE	Cell surface	N.D.	N.D.	Parker et al., 1990a
Bovine coronavirus	HE(truncated)	Secreted	N.D.	N.D.	Parker et al., 1990a
Bovine coronavirus	S peplomer	Cell surface	N.D.	N.D.	Parker et al., 1990b
Mouse hepatitis (JHM)	E2	Cell surface	–	N.D.	Yoden et al., 1989
Hepatitis	S-Ag	Cellular	+	N.D.	Lanford et al., 1989
Hepatitis	S-Ag	Secreted	N.D.	N.D.	Kang et al., 1987
Influenza A	HA	Cell surface	N.D.	N.D.	Possee, 1986
Influenza (fowl plague)	HA	Cell surface	+	Chicken	Kuroda et al., 1986

Virus	Protein	Localization	Neutralizing antibodies[a]	Protection[b]	Reference
Rubella	E1 and E2	N.D.	N.D.	N.D.	Oker-Blom et al., 1989
Measles	F and H	N.D.	N.D.	N.D.	Vialard et al., 1990
Lymphocytic choriomeningitis	GPC	Cell surface	N.D.	N.D.	Matsuura et al., 1986; Matsuura et al., 1987
Hantaan	G1 and G2	N.D.	+	Hamsters	Schmaljohn et al., 1990
Dengue	E	Intracellular	–	N.D.	Zhang et al., 1988
Japanese encephalitis	E	Cell surface	+	N.D.	Matsuura et al., 1989
Rift valley fever	G1 and G2	N.D.	+	Mice	Schmaljohn et al., 1989
Human immunodeficiency	gp160	Cell surface	+	N.D.	Rusche et al., 1987; Wells et al., 1990; Hu et al., 1986; Schwaller et al., 1989; Putney et al., 1987; Farmer et al., 1989; Cochran et al., 1987; Richardson et al., 1988; McQuade et al., 1989
Human immunodeficiency	gp120	Secreted	+	N.D.	Rusche et al., 1987; Wells et al., 1990b; Hu et al., 1986; Schwaller et al., 1989; Putney et al., 1987; Farmer et al., 1989; Cochran et al., 1987; Richardson et al., 1988; McQuade et al., 1989
Human cytomegalovirus	gp55-116 (gB)	N.D.	+	N.D.	Wells et al., 1990a
Herpes simplex	gD	N.D.	+	Mice	Krishna et al., 1989

[a]Virus neutralizing antibodies produced in immunized animals.
[b]Protection of immunized animals from virus challenge.
N.D., not done.

personal communication) even without optimization of a production cell line. In our hands expression levels of TPA in insect cells have never exceeded 1–2 μg/10^6 infected cells. These levels are similar to those reported by other groups (Jarvis and Summers, 1989; Luckow and Summers, 1988b; Furlong et al., 1988; Steiner et al., 1988). TPA was only the second glycoprotein that we expressed in the baculovirus expression system (gp50 was the first), and our knowledge of the system was limited. TPA became a model enzymatically active glycoprotein to evaluate in insect cells.

The TPA cDNA, along with its signal sequences, was inserted in the transfer vector pAC373 (Furlong et al., 1988). The insect cell produced TPA was found to have a lower molecular weight than either CHO or Bowes melanoma produced TPA. This is presumably due to decreased glycosylation, since the carboxyl and amino termini of the insect cell TPA were found to be intact (Furlong et al., 1988). N terminal amino acid analysis also showed that the TPA signal was being cleaved at the appropriate site.

In activity assays, the specific activity of BEVS TPA was found to be the same as, or slightly higher than, Bowes melanoma TPA (Furlong et al., 1988). TPA is known to have enhanced activity in the presence of fibrin of fibrin fragments (Nieuwenhuizen et al., 1983; Ranby, 1982; Verheijen et al., 1982; Hoylaerts et al., 1982). When the activity of TPA from Bowes melanoma and insect cells was compared in the presence of fibrin fragments it was found that insect cell TPA was stimulated twofold less than mammalian TPA. The reasons for this reduction are unknown.

Several other groups have expressed TPA in insect cells. Steiner et al. (1988) found that baculovirus produced TPA was more active than that produced in mouse cells using a bovine papilloma virus (BPV) based vector. Since the latter TPA was more extensively glycosylated, this may indicate that insect cell TPA is more active because of reduced glycosylation. Jarvis and Summers (1989) reported that addition of oligosaccharides was required for secretion of TPA from the cell. They also found that the efficiency of secretion was decreased late in infection. These data suggested that late in the viral infectious cycle the secretory pathways of the cell are compromised. By expressing various forms of the TPA molecule, Devlin et al. (1989) found that the signal peptide was required for TPA to be secreted from infected cells. This provided another example of insect cells recognizing mammalian cell signal sequences, and showed that the secretion of TPA is a specific process rather than simply due to cell lysis.

Since production levels were low compared to mammalian cells, and since BEVS TPA had reduced fibrin enhancement activity, scale-up of the insect cell system was not considered for this protein, though valuable information was gained. It was found that enzymes such as TPA can be expressed in active form in insect cells and that processing is uniform. This information has proven to be important in subsequent expression work.

5.5 EXPRESSION OF A HORMONE AS A THERAPEUTIC AGENT

Nerve growth factor (NGF) is a hormone that stimulates differentiation of adrenergic neurons in the peripheral nervous system (Levi-Montalcine, 1987). It also affects cholinergic neurons in the central nervous system (Kromer, 1987; Whittemore and Seiger, 1987). NGF from only one source has been extensively studied, that from the male mouse submaxillary gland (Bradshaw, 1978; Isackson et al., 1985; Yanker and Shooter, 1982). NGF has been expressed by a vaccinia virus expression vector (Edwards et al., 1988) and was found to be active, but expression levels were not presented. Before expressing NGF in the BEVS system we had attempted to produce it in *E. coli;* NGF was produced but it lacked binding capacity or biological activity (S. Buxser, personal communication).

Our work involved the cloning and expression of the human NGF gene (Buxser et al., 1990). The coding exon from the gene was cloned from human lung fibroblast DNA. When produced naturally, NGF is synthesized as a single polypeptide with a molecular weight of approximately 26 kD. Prepro NGF is then cleaved to the 13 kD mature, active NGF. If BEVS was to be useful as a source of NGF, this processing event would have to occur. The prepro NGF coding sequence was cloned into pVL941 and then transfected into the AcNPV genome.

Figure 4a shows an SDS-PAGE analysis of ^{14}C-glucosamine labeled NGF. Cells were infected with a recombinant virus containing the NGF gene (AcNPV-NGF) and labeled from 24 to 48 hours post infection. The NGF produced in the cells (lane 1) has a molecular weight of 26 kD, which corresponds to the pro NGF form. We were unable to detect a glucosamine labeled protein in the medium (lane 2), though NGF could be purified from the medium. Medium from AcNPV-NGF infected cells was run over anti-NGF monoclonal antibody affinity columns and the purified protein analyzed by SDS-PAGE. The purified NGF shows a single band with a molecular weight of approximately 13 kD, the predicted size of the mature form of NGF. Protein sequence analysis of the amino terminus of the purified NGF showed the amino acid composition

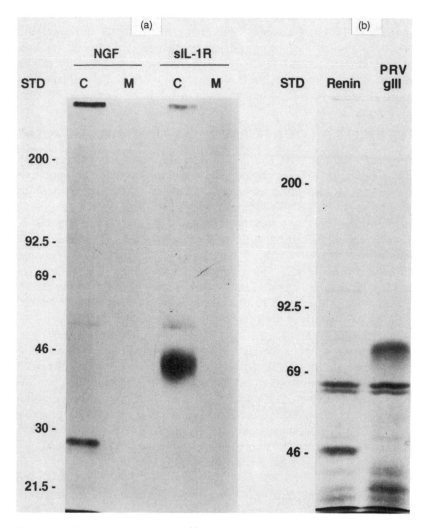

Figure 4 SDS-PAGE analysis of ^{14}C-glucosamine labeled baculovirus infected insect cells. Approximately 3×10^6 Sf-9 cells were infected with the recombinant viruses shown at a MOI of 5 pfu/cell. At 24 hr post infection, medium containing 50 uCi/ml ^{14}C-glucosamine was added to the cells. Label remained on the cells for 24 hr.

a: Cells were infected with recombinant baculoviruses expressing either NGF or soluble interleukin-1 receptor (sIL-1R). Labeled material was separated into cellular (c) and medium (m) fractions prior to electrophoresis.

b: Cells infected with recombinant baculoviruses producing either human renin or PRV gIII. Only the cellular fraction is shown.

predicted from the nucleotide sequence and identical to that found in NGF produced in CHO cells (Iwane et al., 1990).

Glycosylation of the protein was not expected since neither murine nor cobra NGF (Hogue-Angeletti et al., 1976) are glycosylated. However the NGF from two viperid species and two crotalids has been shown to be glycosylated (Pearce et al., 1972; Bailey et al., 1975; Glass and Banthorpe, 1975). It appears from our data that human NGF, at least in the pro NGF form, may be glycosylated. The carbohydrates of the cellular and secreted forms of NGF will have to be analyzed to prove this.

NGF produced in baculovirus infected insect cells was found to be active in both binding and biological activity assays (Buxser et al., 1990). In primary neurite outgrowth assays it was found to stimulate neurite outgrowth and dopamine synthesis in target cells to much the same level as murine NGF. Binding studies showed that insect cell produced NGF bound to nerve growth factor receptor as well as, or slightly better than, purified murine NGF.

NGF has potential therapeutic value. It has been shown that treatment of brain injuries with nerve growth factor can reduce neuronal death (Kromer, 1987), suggesting that NGF could be used to treat severe nervous system injuries. Insect cells produce significant amounts of mature protein, about 2 mg per liter of infected cells. Our results show that this NGF is very similar in its binding and biological activity to mammalian NGF. The ability to produce NGF as needed using the BEV system could solve many of the problems associated with the limited quantities of this protein available for study.

Preliminary crystallization of purified NGF in shown in Figure 5. This crystal is a colorless prism. It is too small (0.15 mm long) for x-ray diffraction analysis, however it does show the potential utility of baculovirus produced proteins for these types of studies.

Other growth factors have been produced in insect cells. Human granulocyte macrophage stimulating factor (Chiou and Wu, 1990) and erythropoietin (EPO) (Quelle et al., 1989; Wojchowski et al., 1987) were found to be glycosylated and secreted after the removal of their signal sequences. Erythropoietin (EPO) produced in insect cells contains reduced glycosylation and no sialic acid and lacks biological activity in vivo. By comparison, CHO cell produced EPO contains sialic acid residues and is active in vivo (Sasaki et al., 1987). It has been reported that asialylation of EPO (purified from serum or urine) abolishes in vivo activity (Tsuda et al., 1990) while increasing in vitro activity. Insect cells may not be the best source for some glycoproteins that require complex oligosaccharide or

Figure 5 Crystal of human nerve growth factor purified from the medium of insect cells infected with a recombinant baculovirus encoding NGF.

sialic acid for their in vivo activity. Although insect cells failed to amidate human gastrin-releasing factor (LeBacq-Verheyden et al., 1988), they did remove the signal sequence from the precursor form of the glycoprotein and secreted an active form of the hormone into the medium of infected cells. Finally, v-sis/platelet-derived growth factor was found to be expressed and secreted at levels 50- to 100-fold higher in insect cells than in mammalian cells (Giese et al., 1989). These data suggest that the baculovirus expression system is an exceptionally good system for the expression of growth factors that do not require complex oligosaccharides or sialic acid for activity.

5.6 EXPRESSION OF A SURFACE RECEPTOR AS A POTENTIAL THERAPEUTIC TARGET

Many mammalian surface receptors have been expressed in the baculovirus expression system: β-adrenergic receptor (George et al., 1989), epidermal growth factor receptor (Greenfield et al., 1988; Patel et al., 1988), glucocorticoid receptor (Srinivasan and Thompson, 1990), various domains of the insulin receptor (Cobb et al., 1989; Ellis et al., 1988; Villalba et al., 1989; Sissom and Ellis, 1989), transferrin receptor (Domingo and Trowbridge, 1988), CD4 (Webb et al., 1989), and CD2 (Alcover et al., 1988). All of these receptors bind to their natural ligands. Soluble forms of CD4 (Hussey et al., 1986; Richardson et al., 1988), epidermal growth factor (Greenfield et al., 1989), nerve growth factor (Vissavajjhala and Ross, 1990), and insulin (Sissom and Ellis, 1989) receptors have also been produced and shown to bind to their natural ligands. In light of this, we felt the BEV system was an appropriate method for expression of the interleukin-1 receptor (IL-1R).

Our goal in producing IL-1R was to produce a reagent that would allow us to test the hypothesis that the removal of active IL-1 in vivo would lead to reduction of chronic inflammation mediated by IL-1. Functionally, IL-1 mediates the immune response to viral and bacterial infections by binding to the IL-1 receptor, which in turn leads to inflammation (Dinarello, 1984). The baculovirus system provided a method to produce large quantities of this receptor, which could then be purified and used to study the interaction of IL-1 with the receptor binding domains. By analyzing these domains it was hoped that we could design drugs which would interfere with this binding. In addition, milligram quantities of pure IL-1 receptor could provide a specific assay tool to facilitate discovery of nonpeptide IL-1 antagonists.

The murine IL-1 receptor is glycoprotein having a molecular weight of 80 kD. The protein is composed of three domains, a 319 amino acid extracellular domain containing seven potential glycosylation sites, a 21 amino acid transmembrane domain, and a 217 amino acid cytoplasmic domain (Sims et al., 1988). To aid in purification, we deleted the regions of the gene encoding the transmembrane and cytoplasmic domains prior to inserting the gene into the transfer vector pVL941. This was expected to yield a soluble form of the receptor which would be secreted into the medium of infected cells.

Soluble IL-1 receptor was produced in infected insect cells (Figure 4a). Only IL-1R in the cell (lane 3) is labeled with ^{14}C-glucosamine; no labeled protein was found in the medium (lane 4). This was somewhat troubling since IL-1 receptor is known to be glycosylated at asparagine residues (Urdal et al., 1988; Bron and MacDonald, 1987). This could represent a slow secretion process, or trimming of labeled carbohydrate from the receptor prior to or during secretion.

Carbohydrate residues were present on the secreted receptor. Soluble IL-1R could be purified from the medium using concanavalin-A Sepharose 4B columns followed by mono-Q chromatography. A single band is seen on SDS-PAGE with a molecular weight of about 44 kD. Western blot analysis showed this band to be IL-1R (data not shown). Amino acid analysis and N terminal sequence analysis of the purified receptor corresponded to the predicted composition and sequence, so the lack of labeled carbohydrate was not due to breakdown of the protein backbone (J. E. McGee, personal communication).

Soluble IL-1R was found to be active in IL-1 binding assays. Activity was measured by Scatchard analysis of radiolabeled IL-1 binding to the purified soluble IL-1R coupled to Sepharose 4B, or to IL-1 receptor from EL4-6.1 cells (MacDonald et al., 1985). K_d values for these two forms of the receptor were very similar. 0.15 +/- 0.02 nM for soluble IL-1R from insect cells versus 0.17 +/- 0.02 nM for EL4-6.1 cell receptor (J. E. McGee, personal communication).

Two methods of producing soluble IL-1R were also tested: shake flasks and a 6 liter airlift fermentor. The medium used in both systems was a commercially available serum-free medium, ExCell 400 (JRH Scientific, Lenexa, Kansas). Cells were grown in the airlift fermentor under conditions based on published procedures (Maiorella et al., 1988). The two methods produced similar levels of protein, from 1 to 2 mg of soluble IL-1 receptor per liter of infected cells (2×10^6 cells/ml). Yields from the airlift fermentor were slightly higher (about 20%), perhaps because of the more accurate control of the growth conditions (oxygen and pH levels) in the fermentor.

Table 3 lists other surface receptors, both full length and truncated (for secretion), that have been expressed in the baculovirus expression system. Of special note are two of the soluble receptors, human insulin receptor and nerve growth factor receptor. Soluble human insulin receptor was found to be secreted slowly, with about half of the molecules in the medium remaining uncleaved. Even though processing was not complete, the portion of the receptor that was cleaved was active and produced at levels 100 times greater than are found in mammalian cells. Soluble nerve growth factor receptor was found to contain little or no carbohydrate and yet it too was active and capable of binding NGF.

In general, even though glycosylation is reduced or absent, receptors expressed in infected insect cells are active and capable of binding their natural ligands. They are also produced at levels high enough to warrant their use for experimental analysis.

Table 3 Expression of Surface Receptors Using the Baculovirus Expression System

	Receptor	Cellular location	Ligand bound	Yield	References
Extracellular domain of receptor					
	Epidermal growth factor	Secreted	EGF	1 mg/10^9 cells	Greenfield et al., 1989
	Nerve growth factor	Secreted	NGF	N.D.	Vissavajjhala and Ross, 1990
	CD4	Secreted	gp120	1–2 mg/10^9 cells	Hussey et al., 1986 Richardson et al., 1988
	Human insulin	Secreted	Insulin	mgs/liter	Sissom and Ellis, 1989
Complete receptor					
	Epidermal growth factor	Cell surface	EGF	1 mg/10^9 cells	Greenfield et al., 1988 Patel et al., 1988
	CD4	Cell surface	N.D.	N.D.	Webb et al., 1989
	β-adrenergic	Cell surface	ICYP DHA	30 nmoles/10^9 cells	George et al., 1989
	Transferrin	Cell surface	Transferrin	5.8 ± 0.9 receptors/cell	Domingo and Trowbridge, 1988
	Glucocorticoid	Cytoplasmic	Glucocorticoids	0.5–1 mg/10^9 cells	Srinivasan and Thompson, 1990
	CD2 (TII)	Cell surface	SRBC	N.D.	Alcover et al., 1988

5.7 EXPRESSION OF AN ENZYME AS A POTENTIAL THERAPEUTIC TARGET

Renin is an aspartyl proteinase whose sole known function is to cleave angiotensinogen (Navar, 1986; Nakanishi et al., 1985; Ondetti and Cushman, 1982). This reaction is the first and rate limiting step in a cascade of factors that regulate arteriole blood pressure (Inagami, 1981). Renin is present in the bloodstream as a zymogen, prorenin, which is activated by a cleavage of the propeptide (Derkx, 1987; Hsueh et al., 1986). The nature of the triggering mechanism that prompts this cleavage is unknown. The absolute specificity of renin for angiotensinogen, as well as its position in the blood pressure regulatory cascade, make it an excellent candidate as a target for specific inhibitors to control hypertension.

The complete coding sequence for human preprorenin was cloned into baculovirus. Figure 4b shows an SDS-PAGE analysis of ^{14}C-glucosamine labeled proteins from recombinant baculovirus infected Sf-9 cells. The infected cells were labeled from 24 to 48 hours post infection. Within the cell, prorenin is made as a single glycoprotein with a molecular weight of about 48 kD. This molecular weight is less than the prorenin produced in transfected CHO cells by about 6 kD (Poorman et al., 1986). This reduction is thought to be due to reduced glycosylation. Prorenin is secreted into the medium; approximately 1 mg is produced per liter of infected cells. After trypsin activation of prorenin from both sources, prorenin from baculovirus infected insect cells was found to be as active as material from transfected CHO cells (R. A. Poorman, personal communication).

The design of inhibitors for an enzyme is greatly facilitated if one can analyze the protein's tertiary structure by x-ray crystallography. X-ray analysis requires large amounts of highly purified, homogeneous protein for study. Figure 6 shows a two dimensional western blot analysis of human prorenin produced and secreted into the medium of insect cells. Only one spot is seen (arrow) that is unique to the renin gel. This spot is homogeneous in size and charge. We attempted this same analysis of CHO produced prorenin but were unable to find any material in the gel which would react to form a spot, even a diffuse one. This may be due to the heterogeneous nature of mammalian produced proteins. These data show that the prorenin produced in infected insect cells is homogeneous and should be suitable for crystallization studies. Of course use of this type of prorenin does not guarantee high resolution crystals.

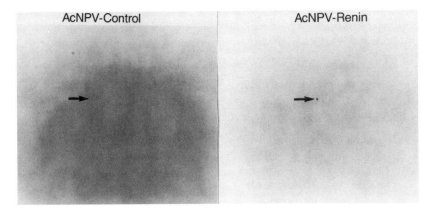

Figure 6 Medium from Sf-9 cells infected with a recombinant baculovirus expressing human renin or a control virus was harvested at 48 hr post infection and analyzed by two-dimensional gel electrophoresis and western blotting. First dimension: ten μl of medium were focused on a 1.5 mm EF tube gel containing broad range ampholytes (pH 3.5–9). Second dimension: the tube gel was layered into a 1.5 mm thick 10–20% polyacrylamide gradient gel and electrophoresed by the method of Laemmli (1970; Davidson et al., 1990; Adams, 1987). Western blotting: proteins from the gel were transferred onto nitrocellulose (Towbin et al., 1979) and probed using a rabbit polyclonal antibody to renin.

CONCLUDING REMARKS

The baculovirus expression system has proven to be a valuable system for the expression of foreign genes. A major advantage of BEVS is the speed with which cloning and expression can be achieved. The time interval from insertion of a gene into a transfer vector to production of milligram quantities of protein for analysis is around six weeks. About three weeks after a transfection, preliminary analysis of the structural integrity, glycosylation, and antigenic structure of the protein can be done.

Analysis of glycoproteins is extremely easy using BEVS. Infected cells can be labeled with glucosamine early in infection, followed by SDS-PAGE analysis of labeled proteins. Since few cellular proteins in insect cells are labeled with glucosamine after infection, the recombinant protein is easy to identify. This is a rapid method to check for structural integrity of the gene as well as for any modifications which might be found on the protein.

Reduced glycosylation found on many insect cell produced glycoproteins has not proved to be a problem for in vitro analysis, at least in our hands. Five of the six genes discussed here encode glycoproteins. The sixth, NGF, might be a glycoprotein. Prorenin, NGF, and TPA were found to be as active as the more heavily glycosylated mammalian proteins. FG and gp50 were immunogenic. There is some evidence that reduced glycosylation of a protein can enhance antigenicity (Alexander and Elder, 1984), which could make baculovirus an ideal expression system for vaccine production. At the same time, enhanced antigenicity may reduce the utility of baculovirus produced proteins as therapeutics. Reduced glycosylation also tends to make a glycoprotein more homogeneous which, in turn, can lead to a more easily analyzed product, perhaps a particular advantage for crystallography.

It was believed that insect cells coult not elongate trimmed N-linked oligosaccharides on glycoproteins to complex side chains containing galactose, fucose, and neuraminic acid (Butters and Hughes, 1981a,b; Hsieh and Robbins, 1984). Recently this has been challenged by several groups. The HA glycoprotein of fowl plague virus was analyzed and found to contain trimmed N-glycans (trimannosyl cores). These cores were further modified with the addition of fucose (Kuroda et al., 1990). Baculovirus expressed TPA was shown to have immature high mannose oligosaccharides removed (Jarvis and Summers, 1989). It is not known if additional carbohydrate was added back to the trimmed high mannose precursor. The analysis of human plasminogen in insect cells provided the first direct evidence of complex carbohydrate addition (Davidson et al., 1990). Human plasminogen was found to contain bisialo-biantennary complex carbohydrates which are identical to those present on the human plasma glycoprotein. These combined data suggest that insect cells have the necessary machinery, such as mannosidases as well as galactosyl-, hexosaminidosyl-, and sialyl-transferases, to process high mannose precursors. It does seem clear, however, that oligosaccharides on proteins produced in insect cells are less complex than on the same protein produced in mammalian cells. The lack of sialic acid on many insect cell derived glycoproteins may have profound effects on in vivo activities.

The baculovirus expression vector system has also been shown to perform many of the posttranslational modifications on heterologous proteins such as: phosphorylation, myristilation, palmitylation, carboxyl methylation, polyisoprenylation, and α-amidation (O'Reilly et al., 1992; Luckow, 1990; Summers, 1989; Luckow and Summers, 1988a; and Summers, 1988). In many cases these modifications have been shown to be inefficient in infected insect cells yielding heterogeneity in the protein product (reviewed in O'Reilly et al., 1992). These deficiencies can cause problems if the protein one wants to express requires posttranslational modifications with true fidelity to the natural protein. Proteins that require extensive modifications for in vivo activity will need to be analyzed extensively prior to use as therapeutics, vaccines, or pharmacologic agents.

The signal and endoproteolytic cleavage, and/or secretion of proteins and glycoproteins in the baculovirus expression system appears to vary depending on the source of the protein (Luckow, 1990; Luckow and Summers, 1988a; O'Reilly et al., 1992; Summers, 1988; Summers, 1989; Miller, 1981; Miller, 1987; Miller et al., 1983; Miller et al., 1986; Kang, 1988; Cameron et al., 1989; Frazer, 1989; Maeda, 1989; Miller, 1988, Miller, 1989). In general proteins which are normally secreted from the cell of origin are also secreted by infected insect cells. Whether a particular endoproteolytic cleavage site is recognized in insect cells must be determined empirically. Secretion and processing of the same protein may be different in different people's hands. Hepatitis B virus surface antigen was produced by two different groups. Lanford et al. (1989) found that 22 nm lipoprotein particles were formed in infected cells but not secreted into the medium, while Kang et al. (1987) reported that 22 nm particles were formed and secreted into the medium.

One final word concerns production levels. In our hands production levels of all the glycoproteins expressed using the BEV system have never consistently exceeded 1 to 3 mg per liter of infected cells (2×10^6 cells/ml). When examined, the use of different transfer vectors based on the polyhedrin promotor had no effect on expression levels. Other researchers have reported the same results for glycoproteins they have produced (Luckow, 1990; Luckow and Summers, 1988a; Summers, 1988; Summers, 1989; Miller, 1981; Miller, 1987; Miller et al., 1983; Miller et al., 1986; Kang, 1988; Cameron et al., 1989; Frazer, 1989; Maeda, 1989; Miller, 1988, Miller, 1989). For analytical purposes these levels are adequate, though it becomes a major concern if one wants to manufacture a low cost product. Future work in our laboratories as well as many others will, it is hoped, increase the efficiency of the system.

In summary, the BEVS is a valuable resource for pharmaceutical research projects to produce vaccines, therapeutic proteins and research tools.

ACKNOWLEDGMENTS

The authors would like to thank the following people for materials and experimental data: T. Yamauchi and K. Murakami for the transfer vector containing the renin gene; D. Chattopadhyay, H. Einspahr, R. Poorman, and J. Hinzman for purification and x-ray crystallographic analysis of NGF; L. Adams for 2-D analysis of renin; S. Sharma for antibody to renin; R. Wardley and P. Berlinski for the PRV animal studies; D. Carter and P. Harris for isolation of the soluble IL-1 receptor; J. Paslay and J. McGee for its biochemical analysis; and K. Hiestand for preparation of the manuscript.

REFERENCES

Adams, L. D. (1987), *Current Protocols in Molecular Biology,* (F. M. Asubel, et al., eds.), John Wiley and Sons, New York, pp. 10.3.1–10.3.12.

Alcover, A., Chang, H. C. Sayre, P. H., Hussey, R. E. and Reinherz, E. L. (1988). *Eur. J. Immunol., 18:*363–368.

Alexander, S. and Elder, J. H. (1984). *Science, 226:*1328–1330.

Bailey, G. S., Banks, B. E. C., Pearce, F. L. and Shipolini, R. A. (1975). *Comp. Biol. Chem. Physiol., 51B:*429–438.

Bailey, M. J., McLeod, D. A., Kang, C-Y. and Bishop, D. H. L. (1989). *Virology, 169:*323–331.

Ben-Porat, T. and Kaplan, A. S. (1970). *Virology, 41:*265–273.

Blissard, G. W. and Rohrmann, G. F. (1990). *Ann. Rev. Entomol., 35:*127–155.

Bradshaw, R. A. (1978). *Ann Rev. Biochem., 47:*191–216.

Brideau, R. J., Walters, R. R., Stier, M. A. and Wathen, M. W. (1989). *J. Gen. Virol., 70:*2637–2644.

Bron, C. and MacDonald, H. R. (1987). *FEBS Lett., 219:*365–368.

Browne, M. J., Dodd, I., Carey, J. E., Chapman, C. G. and Robinson, J. H. (1985). *Thromb. Haemostasis, 54:*422–424.

Butters, T. D. and Hughes, R. C. (1981a). *Biochem. Biophys. Acta, 640:*655–671.

Butters, T. D. and Hughes, R. C. (1981b). *Biochem. Biophys. Acta, 640:*672–686.

Buxser, S., Vroegop, S., Decker, D., Hinzmann, J., Poorman, R. A., Thomsen, D. R., Stier, M. A., Abraham, I., Greenberg, B. D., Hatzenbuhler, N. Y., Shea, M., Curry, K. A. and Tomich, C-S. C. (1991). *J. Neurochem.,* Vol. 56: 1012–1018.

Cameron, I. R., Possee, R. D. and Bishop, D. H. L. (1989). *Trends Biotech.,* 7:66–70.

Chiou, C-J. and Wu, M-C. (1990). *FEBS Lett., 259:*249–253.

Christman, J. K., Silverstein, S. C. and Acs, G. (1977). *Proteinases in Mammalian Cells and Tissues,* (A. J. Barret, ed.), Elsevier, Amsterdam.

Cobb, M. H., Sang, B-C., Gonzalez, R., Goldsmith, E. and Ellis, L. (1989). *J. Biol. Chem., 264:*18701–18706.

Cochran, M. A., Ericson, B. L., Knell, J. D. and Smith, G. E. (1987). *Vaccines 87,* (R. M. Chanock, Lerner, R. A. and Brown, F, eds.), Cold Spring Harbor Laboratory, Cold Spring Harbor, New York.

Collen, D. J. (1980). *Thromb. Haemostasis, 43:*77–89.

Collen, D. and Lunen, H. R. (1984). *Arteriosclerosis, 4:*579–585.

Collen, D., Stassen, J. M., Marofina, B., Builder, S. and DeCock, F. (1984). *J. Pharmacol. Exp. Ther., 231:*146–152.

Davidson, D. J., Fraser, M. J. and Castellino, F. J. (1990). *Biochem., 29:*5584–5590.

Derkx, F. H. M. (1987). *Human Prorenin,* ICG Printing, Dordrecht.

Devlin, J. J., Devlin, P. E., Clark, R., O'Rourke, E. C., Levenson, C. and Mark, D. F. (1989). *Bio/Technology, 7:*286–292.

Dinarello, C. A. (1984). *Rev. Inf. Dis., 6:*51–95.

Doerfler, W. and Bohm, P. (1986). The molecular biology of baculoviruses, *Curr. Top. Microbiol. Immunol., 131.* Springer-Verlag, New York.

Domingo, D. L. and Trowbridge, I. S. (1988). *J. Biol. Chem., 263:*13386–13392.

Dubovi, E. J. (1982), *J. Virol., 42:*372–378.

Edwards, R. H., Selby, M. J., Mobley, W. C., Weinrich, S. L., Hruby, D. E. and Rutter, W. J. (1988). *Mol. Cell. Biol., 8:*2456–2464.

Ellis, L., Levitan, A., Cobb, M. H. and Ramos, P. (1988). *J. Virol., 62:*1634–1639.

Farmer, J. L., Hampton, R. G. and Boots, E. (1989). *J. Virol. Meth., 26:*279–290.

Faulkner, P. (1981). *Pathogenesis of Invertebrate Microbial Diseases,* (E. W. Davidson, ed.), Allanheld-Osmund, Totawa, New Jersey, pp. 3–37.

Frazer, M. J. (1989). *In vitro Cell. Dev. Biol., 25:*225–235.

Furlong, A. M., Thomsen, D. R., Marotti, K. R., Post, L. E. and Sharma, S. K. (1988). *Biotechnol. Appl. Biochem., 10:*459–464.

George, S. T., Arbabian, M. A., Ruoho, A. E., Kiely, J., and Molbon, C. C. (1989). *Biochem. Biophys. Res. Comm., 163:*1265–1269.

Giese, N., May-Siroff, M., LaRochelle, W. J., Van Wyke-Coelingh, K. and Aaronson, S. A. (1989). *J. Virol., 63:*3080–3086.

Glass, R. E. and Banthorpe, D. V. (1975). *Biochem. Biophys. Acta, 405:*23–26.

Granados, R. R. and Federici, B. A. (1986a), *The Biology of Baculoviruses, Vol. I: Biological Properties and Molecular Biology,* CRC Press, Boca Raton, Florida.

Granados, R. R. and Federici, B. A. (1986b). *The Biology of Baculoviruses, Vol. 2: Practical Applications for Insect Control,* CRC Press, Boca Raton, Florida.

Greenfield, C., Patel, G., Clark, S., Jones, N. and Waterfield, M. D. (1988). *EMBO J., 7:*139–146.

Greenfield, C., Hiles, I., Waterfield, M. D., Federwisch, M., Wollmer, A., Blundell, T. L. and McDonald, N. (1989). *EMBO J., 8:*4115–4124.

Gruber, C. and Levine, S. (1983). *J. Gen. Virol., 64:*825–832.

Gustafson, D. P. (1981). *Comparative Diagnosis of Viral Diseases, Vol. 3.* (E. Kurstak and C. Kurstak, eds.), Academic Press, New York.

Hink, W. G. (1970). *Nature, 226:*466–467.

Hogue-Angeletti, R. A., Frazier, W. A., Jacobs, J. W., Niall, H. D., and Bradshaw, R. A. (1976). *Biochem., 15:*26–34.

Hoylaerts, M., Rijken, D. C., Lijnen, H. R. and Collen, D. (1982). *J. Biol. Chem., 257:*2912–2919.

Hsieh, P. and Robbins, P. W. (1984). *J. Biol. Chem., 259:*2375–2382.

Hsueh, W. A., Do, Y. S., Shinagawa, T., Tam, H., Ponte, P. A., Baxter, J. D., Shine, J. and Fritz, L. C. (1986). *Hypertension, 8:*78–83.

Hu, S-K., Kosowski, S. G. and Schaaf, K. F. (1986). *J. Virol., 61:*3617–3620.

Huang, Y. T., Collins, P. L. and Wertz, G. W. (1985). *Virus Res., 2:*157–173.

Hussey, R. E., Richardson, N. E., Kowalski, M., Brown, N. R., Chang, H. C., Siliciano, R. F., Dorfman, T., Walker, B., Sodroski, J. and Reinherz, E. L. (1986), *Nature, 331:*78–81.

Inagami, T. (1981). *Biochemical Regulation of Blood Pressure* (R. L. Soffer, ed.), John Wiley and Sons, New York.

Inlow, D. A., Shauger, A. and Maiorella, B. (1989). *J. Tissue Culture Meth., 12:*13–16.

Isackson, P. J., Dunbar, J. C. and Bradshaw, R. A. (1985). *Intl. J. Neuroscience, 26:*95–108.

Iwane, M., Kitamura, Y., Kaisho, Y., Yoshimura, K., Shintani, A., Sasada, R., Nakagawa, S., Kawahara, K., Nakahama, K. and Kakinuma, A. (1990). *Biochem. Biophys. Res. Comm., 171:*161–162.

Jarvis, D. L. and Summers, M. D. (1989). *Mol. Cell. Biol., 9:*214–223.

Johnson, D. C. and Ligas, M. W. (1988). *J. Virol., 62:*4605–4612.

Kang, C-Y., Bishop, D. H. L., Seo, J-S., Matsuura, Y, and Choe, M. (1987). *J. Gen. Virol. 68:*2607–2613.

Kang, C-Y., (1988). *Adv. Virus Res., 35:*177–192.

Kaufman, R., Wasley, L. C., Spiliotes, A. J., Gossels, S. D., Lat, S. A., Larsen, G. R. and Kay, R. M. (1985). *Mol. Cell. Biol., 5:*1750–1759.

Kim, H. W., Canchola, J., Brandt, C., Pyles, G., Chanock, R. M., Jenson, K. and Parrott, R. H. (1969). *Am. J. Epidemiol., 89:*422–434.

Kingsbury, D. W., Bratt, M. A., Choppin, P. W., Hansen, R. P., Hosaka, Y., Ter Meulen, V., Norrby, E., Plowright, W., Rott, R. and Wunner, W. H. (1978). *Intervirology, 10:*137–152.

Koener, J. F. and Leong, J-A. C. (1990). *J. Virol., 64:*428–430.

Krishna, S., Blacklaws, B. A., Overton, H. A., Bishop, D. H. L. and Nash, A. A. (1989). *J. Gen. Virol., 70:*1805–1814.

Kromer, L. F. (1987). *Science, 235:*214–216.

Kuroda, K., Hauser, C., Rott, R., Klenk, H-D. and Doerfler, W. (1986). *EMBO J., 5:*1359–1365.

Kuroda, K., Geyer, H., Geyer, R., Doerfler, W. and Klenk, H-D. (1990). *Virology, 174:*418–429.

Laemmli, U. K. (1970). *Nature, 227:*600–685.

Lambert, D. (1988). *Virology, 164:*458–466.

Lanford, R. E., Luckow, V., Kennedy, R. C., Dressman, G. R., Notvall, L. and Summers, M. D. (1989). *J. Virol.*, *63:*1549–1557.

LeBacq-Verheyden, A-M., Kasprzyk, P. G., Raum, M. G., Van Wyke-Coelingh, K., LeBacq, J. A. and Battey, J. F. (1988). *Mol. Cell. Biol.*, *8:*3129—3135.

LeMontt, J. F., Wei, C. M. and Dackowski, W. R. (1985). *DNA, 4:*419–428.

Levi-Montalcine, R. (1987). *Science, 237:*1154–1162.

Levine, S., Klaiber-Franco, R. and Paradiso, P. R. (1987). *J. Gen. Virol., 68:*2521–2524.

Levine, S., Dajani, A. and Klaiber-Franco, R. (1988). *J. Gen. Virol., 69:*1229–1239.

Luckow, V. A. and Summers, M. D. (1988a). *Bio/Technology, 6:*47–55.

Luckow, V. A. and Summers, M. D. (1988b). *Virology, 167:*56–71.

Luckow, V. A. and Summers, M. D. (1989). *Virology, 170:*31–39.

Luckow, V. A. (1990). *Recombinant DNA Technology and Applications* (C. Ho, A. Prokop, and R. Bajpai, eds.), McGraw-Hill, New York.

MacDonald, H. R., Lees, R. K. and Bran, C. (1985). *J. Immunol., 135:*3944–3950.

MacKay, V. L. (1988). *Biochemistry and Molecular Biology of Industrial Yeasts* (G. G. Stewart and R. D. Klein, eds.), CRC Press, Boca Raton, Florida.

Maeda, S. (1989). *Ann Rev. Entomol., 34:*351–372.

Maiorella, B., Inlow, D., Shauger, A. and Harano, D. (1988). *Bio/Technol., 6:*1406–1410.

Marchioli, C. C., Yancey, R. J., Petrovskis, E. A., Timmins, J. G. and Post, L. E. (1987), *J. Virol., 61:*3977–3982.

Matsuura, Y., Possee, R. D. and Bishop, D. H. L. (1986), *J. Gen. Virol., 67:*1515–1529.

Matsuura, Y., Possee, R. D., Overton, H. A. and Bishop, D. H. L. (1987). *J. Gen. Virol., 68:*1233–1250.

Matsuura, Y., Miyamoto, M., Sato, T., Morita, C. and Yasui, K. (1989). *Virology, 173:*674–682.

McQuade, T. J., Pitts, T. W. and Tarpley, W. G. (1989). *Biochem. Biophys. Res. Comm., 163:*172–176.

Miller, L. K. (1981). *Genetic Engineering in Plant Sciences* (N. J. Panopoulas, ed.), Praeger, New York, pp. 203–235.

Miller, L. K., Miller, D. W. and Adang, M. J. (1983). *Genetic Engineering in Eukaryotes* (P. F. Lutquin and A. Kleinhofs, eds.), Plenum Press, New York, pp. 89–98.

Miller, D. W., Safer, P. and Miller, L. K. (1986). *Genetic Engineering* (J. K. Setlow and A. Hollaender, eds.), Plenum Press, New York, vol. 8 pp. 277.

Miller, L. K. (1987). *Vectors: A Survey of Molecular Cloning Vectors and Their Uses* (D. Denhardt and F. Rodriguez, eds.), Butterworth, Stoneham, Massachusetts, pp. 457–465.

Miller, L. K. (1988). *Ann. Rev. Microbiol. 42:*177–199.

Miller, L. K. (1989). *Bioessays, 11:*91–95.

Murphy, B. R., Prince, G. A., Walsh, E. E., Kim, H. W., Parrott, R. H., Hemming, V. G., Rodriguez, W. J. and Chanock, R. M. (1986). *J. Clin. Microbiol. 24:*197–202.

Nakanishi, S., Kitamura, N. and Ohkubo, H. (1985). *Bio/Technol., 3:*1089–1098.

Navar, L. G. (1986). *Symposium: Fed. Proc., 45:*1411–1453.

Nieuwenhuizen, W., Verheijen, J. H., Vermond, A. and Chang, G. T. G. (1983). *Biochem. Biophys. Acta., 755:*531–533.

Oker-Blom, C., Pettersson, R. F. and Summers, M. D. (1989). *Virology, 172:*82–91.

Olmsted, R. A., Murphy, B. R., Lawrence, L. A., Elango, N., Moss, B. and Collins, P. L. (1989). *J. Virol., 63:*411–420.

Ondetti, M. A. and Cushman, D. W. (1982). *Ann. Rev. Biochem., 51:*283–309.

O'Reilly, D. R., Miller, L. K., and Luckow, V. A. (1992). *Baculovirus Expression Vectors. A Laboratory Manual,* W.H. Freeman and Co., New York.

Parker, M. D., Yoo, D. and Babiuk, L. A. (1990a). *J. Virol., 64:*1625–1629.

Parker, M. D., Yoo, D., Cox, G. J. and Babiuk, L. A. (1990b). *J. Gen. Virol., 71:*263–270.

Patel, G., Greenfield, C., Stabel, S., Waterfield, M. D., Parker, P. J. and Jones, N. C. (1988). *Current Communications in Molecular Biology: Viral Vectors.* (Y. Gluzman and S. H. Hughes, eds.), Cold Spring Harbor Laboratory, Cold Spring Harbor, New York.

Pearce, F. L., Banks, B. E. C., Banthorpe, D. V., Berry, A. R., Davies, H. S. and Vernon, C. A. (1972), *Eur. J. Biochem., 29:*417–425.

Peeples, M. and Levine, S. (1979). *Virology, 95:*137–145.

Pennica, D., Holmes, W. E., Kohr, W. J., Hawkins, R. N., Vehr, G. A., Ward, C. A., Bennett, W. F., Yelverton, E., Seeburg, P. H., Heyneker, H. L. and Beoddel, D. V. (1983). *Nature, 301:*214–221.

Petrovskis, E. A., Timmins, J. G., Armentrout, M. A., Marchioli, C. C., Yancey, R. J. and Post, L. E. (1986). *J. Virol., 59:*216–223.

Poorman, R. A., Palermo, D. P., Post, L. E., Murakami, K., Kinner, J. H., Smith, C. W., Reardon, I. and Heinrikson, R. L. (1986). *Proteins, 1:*139–145.

Possee, R. D. (1986). *Virus Res., 5:*43–59.

Prehaud, C., Takehara, K., Flamand, A. and Bishop, D. H. L. (1989). *Virology, 173:*390–399.

Putney, S., Rusche, J. Matthews, T., Krohn, K., Carson, H., Lynn, D., Jackson, J., Robey, W. G. and Ranki, A. (1987). *UCLA Symposia on Molecular and Cellular Biology, New Series* (D. Bolognesi, ed.), University of California, Los Angeles, California.

Quelle, F. W., Caslake, L. F., Burkert, R. E. and Wojchowski, D. M. (1989). *Blood, 74:*652–657.

Randy, M. (1982), *Biochem. Biophys. Acta, 704:*461–469.

Ray, R., Galinski, M. S. and Compans, R. W. (1989). *Virus Res., 12:*169–180.

Richardson, N. E., Brown, N. R., Hussey, R. E., Vaid, A., Matthews, T. J., Bolognesi, D. P. and Reinherz, E. L. (1988). *Proc. Natl. Acad. Sci. U.S.A., 85:*6102–6106.

Rijken, D. C. and Collen, D. (1981). *J. Biol. Chem.*, *256*:7035–7041.

Rusche, J. R., Lynn, D. L., Robert-Guroff, M., Langlois, A. J., Lyerly, H. K., Carson, H., Krohn, K., Ranki, A., Gallo, R. C., Bolognesi, D. P., Putney, S. D. and Matthews, T. J. (1987). *Proc. Natl. Acad. Sci. U.S.A.*, *84*:6924–6928.

Sambrook, J., Hanahan, D., Rodgers, L. and Gething, M. J. (1986). *Mol. Biol. Med.*, *3*:459–481.

Sasaki, H., Bothner, B., Dell, A. and Fukuda, M. (1987). *J. Biol. Chem.*, *262*:12059–12076.

Schmaljohn, C. S., Parker, M. D., Ennis, W. H., Dalrymple, J. M., Collett, M. S., Suzich, J. A. and Schmaljohn, A. L. (1989). *Virology, 170*:184–192.

Schmaljohn, C. S., Chu, Y-K., Schmaljohn, A. L. and Dalrymple, J. M. (1990). *J. Virol.*, *64*:3162–3170.

Schwaller, M., Smith, G. E., Skehel, J. J. and Wiley, D. C. (1989). *Virology, 172*:367–369.

Sims, J. E., March, C. J., Cosman, D., Widmer, M. B., MacDonald, H. R., McMahan, C. J., Grubin, C. E., Wignall, J. M., Jackson, J. L., Call, S. M., Friend, D., Alpert, A. R., Gillis, S., Urdal, D. L. and Dower, S. K. (1988). *Science, 241*:585–588.

Sissom, J. and Ellis, L. (1989). *Biochem. J., 261*:119–126.

Smith, G. E., Summers, M. D. and Fraser, M. J. (1983). *Mol. Cell. Biol., 3*:2156–2165.

Smith, G. E., Ju, G., Ericson, B. L., Moschera, J., Lahm, H., Chizzonite, R. and Summers, M. D. (1985), *Proc. Natl. Acad. Sci. U.S.A., 82*:8404–8408.

Srinivasan, G. and Thompson, E. B. (1990). *Mol. Endocrinol., 4*:209–216.

Steiner, H., Pohl, G., Gunne, H., Hellers, M., Elhammer, A. and Hansson, L. (1988). *Gene, 73*:449–458.

Summers, M. D. (1978). *Atlas of Insect and Plant Viruses* (K. Maramarasch, ed.), Academic Press, New York, pp. 3–33.

Summers, M. D. and Smith, G. E. (1987). Texas Agricultural Experimental Station Bulletin No. 1555.

Summers, M. D. (1988). *Current Communications in Molecular Biology: Viral Vectors* (Y. Gluzman and S. H. Hughes, eds.), Cold Spring Harbor Laboratory, Cold Spring Harbor, New York, pp. 91–97.

Summers, M. D. (1989). *Concepts in Viral Pathogenesis* (A. L. Notkins and M. B. A. Oldstone, eds.), Springer-Verlag, New York, pp. 77–86.

Tanada, Y. and Hess, R. T. (1984), *Insect Ultrastructure, Vol. 2* (R. C. King and H. Akai, eds.), Plenum Press, New York, pp. 517–556.

Thomsen, D. R., Post, L. E. and Elhammer, A. P. (1990). *J. Cell Biochem., 43*:67–79.

Towbin, H., Staehlin, T., and Gordon, J. (1979). *Proc. Natl. Acad. Sci. U.S.A., 76*:4350–4354.

Tsuda, E., Kawanishi, G., Ueda, M., Masuda, S. and Sasaki, R. (1990). *Eur. J Biochem., 188*:405–411.

Urdal, D. L., Call, S. M., Jackson, J. L. and Dower, S. K. (1988), *J. Biol. Chem., 263*:2870–2877.

Van Wyke-Coelingh, K. L., Murphy, B. R., Collins, P. L., LeBacq-Verheyden, A. M. and Battey, J. F. (1987). *Virology, 160:*465–472.

Verheijen, J. H., Nieuwenhuizen, W. and Wijngaards, G. (1982). *Throm. Res., 27:*77–385.

Vialard, J., Lalumiere, M., Vernet, T., Briedis, D., Alkhatib, G., Henning, D., Levin, D. and Richardson, C. (1990). *J. Virol., 64:*37–50.

Villalba, M., Wente, S. R., Russell, D. S., Ahn, J., Reichelderfer, C. F. and Rosen, O. M. (1989). *Proc. Natl. Acad. Sci. U.S.A., 86:*7848–7852.

Vissavajjhala, P. and Ross, A. H. (1990). *J. Biol. Chem., 265:*4746–4752.

Wallen, P., Pohl, G., Bergsdorf, N., Ranby, M., Ny, T. and Jornvall, H. (1983). *Eur. J. Biochem., 132:*681–686.

Walsh, E. E. and Hruska, J. (1983). *J. Virol., 47:*171–177.

Wathen, M. W. and Wathen, L. M. K. (1984), *J. Virol., 51:*57–62.

Wathen, M. W., Brideau, R. J. and Thomsen, D. R. (1989a). *J. Inf. Diseases, 159:*255–264.

Wathen, M. W., Brideau, R. J., Thomsen, D. R. and Murphy, B. R. (1989b). *J. Gen. Virol., 70:*2625–2635.

Webb, N. R., Madoulet, C., Tossi, P. F., Broussard, D. R., Sneed, L., Nicholau, C. and Summers, M. D. (1989). *Proc. Natl. Acad. Sci. U.S.A., 86:*7731–7735.

Wells, D. E. and Compans, R. W. (1990a). *Virology, 176:*575–586.

Wells, D. E., Vugler, L. G. and Britt, W. J. (1990b). *J. Gen. Virol., 71:*873–880.

Wertz, G. W., Collins, P. L., Huang, Y, Gruber, C., Levine, S. and Ball, L. A. (1985). *Proc. Natl. Acad. Sci. U.S.A., 82:*4075–4079.

Whittemore, S. R. and Seiger, A. (1987). *Brain Res. Rev., 12:*439–464.

Wojchowski, D. M., Orkin, S. A., Sytkowski, A. J. (1987). *Biochem. Biophys. Acta, 910:*224–232.

Yanker, B. A. and Shooter, E. M. (1982), *Ann. Rev. Biochem., 51:*845–868.

Yoden, S., Kikuchi, T., Siddell, S. G. and Taguchi, F. (1989). *Virology, 173:*615–623.

Zhang, Y-M., Hayes, E. P., McCarty, T. C., Dubois, D. R., Summers, P. L., Eckels, K. H., Chanock, R. M. and Lai, C-J. (1988). *J. Virol., 62:*3027–3031.

6

Scale-Up Considerations and Bioreactor Development for Animal Cell Cultivation

Johannes Tramper, Kees D. de Gooijer, and Just M. Vlak
Wageningen Agricultural University, Wageningen, The Netherlands

6.1 INTRODUCTION

Background

Any cell, when placed in a moving fluid with velocity gradients, experiences a shear force, the magnitude of which depends on the dynamic viscosity of the fluid, the fluid velocity gradients, and the size of the cell. The effects of this shear force largely depend on the properties of the cell itself. In comparison to microorganisms, insect cells, and animal cells in general, are very fragile. This is the result of their relatively large size and the lack of a cell wall. In larger bioreactors in particular, achieving an adequate oxygen supply is hampered by this cell fragility. Special measures are thus required for fragile cell bioreactors, and these will be discussed here.

Strictly speaking, shear forces result from spatial differences in the levels of momentum across material stream lines in a moving body of fluid. In a stirred bioreactor, however, cells can encounter a variety of other mechanical forces due to collisions with the vessel walls, the agitator, or other objects in the bioreactor. In addition, sparged gas bubbles subject the cell to surface tension forces and to fluid mechanical forces resulting from the motion, disengagement and bursting of bubbles, and from foaming. Here, all will be collectively referred to as shear, and if possible quantified in terms of shear rate, shear stress, or smallest turbulent eddy length.

Shear

The relation between the shear stress τ (N · m^{-2}), the dynamic viscosity η (N · s · m^{-2}) and the shear rate $\dot{\gamma}$ (s^{-1}) is given by Newton's equation

$$\tau = -\eta \frac{dv}{dx} = \eta \dot{\gamma} \tag{1}$$

with dv/dx(s^{-1}) the fluid velocity gradient. In Figure 1 the local fluid velocity v (m/s) is plotted as a function of the local coordinate x (m). According to this figure and to Equation (1), the shear stress is assumed to be positive in the direction of decreasing fluid velocity. The dynamic viscosity of real Newtonian fluids is a constant, dependent only on pressure and temperature. Most biotechnological fluids are dilute aqueous media for which Newton's equation is appropriate. Furthermore, from a practical and engineering point of view, the existing relations for non-Newtonian behavior, e.g., in case of very thick cell suspensions, are not very suitable for describing fluid flow in technical equipment. Therefore, only Newtonian fluids will be considered here.

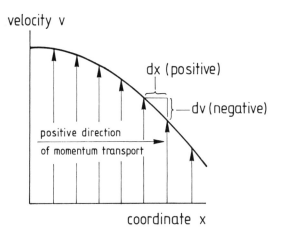

Figure 1 Explanation of shear stress in fluids (from Van 't Riet and Tramper, 1991).

Fluid Flow

Fluid flow can basically be divided into two types, laminar flow and turbulent flow. In laminar flow, fluid elements move along parallel stream lines. In this case shear stresses in the fluid are predictable from velocity gradients. This is not possible in turbulent flow. Hinze (1959) defines turbulent flow as follows: "Turbulent fluid motion is an irregular condition of flow in which the various quantities show a random variation with time and space coordinates, so that statistically distinct average values can be discerned." For instance, time-average shear stress, used later on, refers to this situation. Turbulence can be generated by friction forces at solid objects, e.g., impellers, or by the flow of layers of fluid with different velocities past or over one another, for example as a result of air bubbles moving through the liquid. As turbulence is a common situation in bioreactors, much attention will be paid to this aspect, especially since turbulence can be lethal to fragile cells.

Shear Sensitivity

For the rational design and scale-up of bioreactors, the shear sensitivity of fragile cells should be described in quantitative terms. Since there is relatively little information of this kind available in the literature, experiments are usually the only way to collect quantitative data on the influence of shear on a particular cell type. Although these experiments may be rather difficult in many cases, the general strategy is clear (Märkl and Bronnenmeier, 1985). The key parameter measured for quantification of shear effects should be closely related to the aim of the technical process under investigation, for instance cell viability and growth, formation rate or quality of the desired product, yield coefficients, overall productivity, etc. Secondary parameters, such as release of intracellular material or microscopically obtained morphological data may be very helpful, but they can be misleading when their relation to the direct process aim is not incorporated into the investigation. Furthermore, it is important to separate the influence of damage due to shear and damage due to transport effects, such as mass and heat transfer limitations.

Meijer (1989) expresses the opinion that shear sensitivity data should preferably be collected in a down-scaled version of the intended production system. Therefore, if it is the intention to develop a large-scale process based on an impeller stirred tank, the recommended method would be to collect data in a small impeller stirred tank. Meijer recognizes the disadvantage of the poorly defined and irregular shear levels in such stirred vessels, but, on the other hand, data collected in a device with well-defined and constant shear levels usually requires an awkward translation to the practical situation. Probably the best approach is to analyse the available shear-determining devices and damage-measuring methods for each particular cell type and process aim before making a choice. Here the emphasis is put on stirred vessels, bubble columns and airlift loop reactors, as these are in general the bioreactors of choice for larger scale productions, and also for fragile cells, despite the relatively high levels of shear. These well-mixed bioreactors have a variety of advantages such as scaleability, ease of controlling and monitoring important bioreactor parameters, relatively uniform bioreactor conditions, and use of existing industrial capacity and experience from other biological processes (Kunas and Papoutsakis, 1990). Understanding the mechanisms of damage caused by fluid mechanical forces associated with mixing and/or aeration is, however, a prerequisite for adequate bioreactor design.

6.2 THE STIRRED VESSEL

Introduction

The standard fermentor has been and still is the workhorse of the bioreactor stable used in biotechnology. Consequently, the practical experience with these stirred vessels is enormous and so is the desire to use them, even for fragile cells, despite the obvious disadvantage of poorly defined, irregular, and often high shear levels. Many studies aimed at determining the fragility of cells have therefore been executed in stirred vessels (Van 't Riet and Tramper, 1991 and references cited therein). A detailed analysis of the hydrodynamic effects on anchorage-dependent animal cells attached to microcarriers in stirred vessels can be found in Cherry and Papoutsakis (1986) and Croughan et al. (1987 and 1989). Animal cells on microcarriers are especially susceptible to shear. In addition to the lack of a protective cell wall and their relatively large size (diameter of about 15 μm), they also lack individual cell mobility. Attached cells thus cannot freely rotate or translate, and accordingly can not reduce the net forces and torques experienced in the shear fields of the moving fluids in a stirred vessel.

The purpose of stirring the medium in a bioreactor is threefold. First, it is required to prevent the settling of cells, and second, to assure a homogeneous environment for the cells, i.e., a continuous and adequate supply of nutrients. Finally, stirring is used to improve the oxygen transfer from the gas to the liquid phase. In order to reach these goals, certainly the latter two, the stirrer speed usually needs to be so high that the fluid flow will be turbulent. Analysis of turbulent flow fields with respect to shear is therefore pivotal and will be discussed in some detail.

Integrated Shear Factor

Croughan et al. (1987) correlate the growth of anchorage dependent animal cells, among others, with an integrated shear factor (ISF; s^{-1}) as given by

$$\text{ISF} = 2\pi N \frac{D}{T_v - D} \tag{2}$$

with N the stirrer speed (s^{-1}), D the stirrer diameter (m), and T_v the vessel diameter (m). The integrated shear factor is a measure of the strength of the shear field between the impeller and vessel wall. They found that above an ISF of about 20 s^{-1} the growth of FS-4 cells on microcarriers rapidly drops (Figure 2). This figure shows one set of data for stirred vessels with volumes of 0.25 and 2 liters, and five different impeller diameters. Although the ISF may thus have some use with respect to scale-up, especially as this factor is very easy to calculate, much more work is needed, in particular in larger vessels, to establish the real usefulness of this factor for rational design and scale-up of stirred vessels for growth of fragile cells in more general terms. For

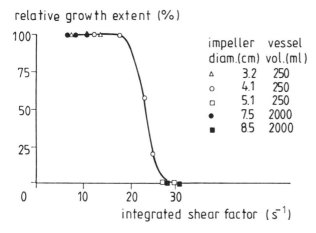

Figure 2 Relative growth extent of FS-4 cells attached to microcarriers as a function of the integrated shear factor (adapted from Croughan et al., 1987).

insect cells in suspension, Tramper et al. (1986) have found that death rapidly occurs at a stirrer speed of about $9 \, s^{-1}$ in a 1 liter round bottomed fermenter ($T_v = 9$ cm) equipped with a marine impeller ($D = 4$ cm). If these values are substituted in Equation (2), a critical ISF of $45 \, s^{-1}$ is calculated. As the data relate to cells in suspension, a higher value than found for cells attached to microcarriers can be expected, for reasons given above.

Time-Averaged Shear Rate

Croughan et al. (1987) also correlate the growth of animal cells on microcarriers with the time-averaged shear rate $\dot{\gamma}_{ave}(s^{-1})$ in the region of the reactor with a radius larger than the stirrer radius. For an unbaffled vessel operated in the turbulent regime ($\text{Re} > 1000$) it can be derived from their equations that

$$\dot{\gamma} = \frac{113.1 ND^{1.8}(T_v^{0.2} - D^{0.2}) \dfrac{0.625 \, \text{Re}}{625 + \text{Re}}}{T_v^2 - D^2} \tag{3}$$

with Re the Reynolds number:

$$\text{Re} = ND^2 \frac{\rho}{\eta} \tag{4}$$

where ρ is the liquid density ($kg \cdot m^{-3}$). The same data on relative growth of FS-4 cells on microcarriers, Figure 2, were plotted by these authors as a function of this average shear rate as calculated from Equation (3). Similar to the ISF, a sharp drop in growth occurred at a certain point, in this case at a critical $\dot{\gamma}_{ave}$ of about $2.5 \, s^{-1}$. Similar results were also found for the maximum cell concentration of chicken embryo fibroblasts grown on microcarriers with a critical $\dot{\gamma}_{ave}$ of about $7 \, s^{-1}$. This similarity to ISF can be expected for geometrically similar systems operated at high Reynolds numbers ($\text{Re} \gg 625$). Both ISF and $\dot{\gamma}_{ave}$ are then proportional to N only, with Equations (2) and (3) reducing to

$$\dot{\gamma}_{ave} = cN \tag{5}$$

where the constant c takes on a value of 2π and 3.5, respectively, for a D/T_v ratio of 0.5. This is the same form of equation as for the shear rate in the impeller region for fluids agitated in the laminar regime (Metzner et al., 1961):

$$\dot{\gamma} = 10N \tag{6}$$

Table 1 Experimental Values of c for
Pseudoplastic Fluids

Impeller type	c
Curved blade paddle	7
Paddle	10–13
Six blade turbine	10–13
Propellor	10
Anchor	20–25
Helical ribbon	30

Source: Margaritis and Zajic, 1978.

Also according to Calderbank and Moo-Young (in Atkinson and Mavi-tuna, 1983) the average shear rate in the impeller region of a stirred tank for the laminar flow regime (Re < 10) can be expressed as Equation (5) with c as a constant independent of the rheological characteristics of the fluid, but dependent on the measuring system used. Table 1 gives some values for c for pseudoplastic fluids, which thus also holds for other type of fluids when measured in the same vessel.

Calculation of $\dot{\gamma}_{ave}$ at the critical stirrer speed for insect cells (see above) with Equations (3) and (4) yields a value of 19 s^{-1}, which is, as can be expected, higher than the values for the FS-4 cells and chicken embryo fibroblasts on microcarriers. To calculate Re, a liquid density ρ of 1000 kg · m^{-3} was substituted, and a dynamic viscosity η of 10^{-2} N · s · m^{-2}, as our insect cell media for reactors routinely contain 0.1% methylcellulose. Re so calculated is 1440, i.e., in the turbulent regime. Another explanation for the somewhat lower fragility of insect cells in suspension compared to cells attached to microcarriers could be, as will be discussed later, the higher viscosity of the insect cell medium.

Maximum (Time-Averaged) Shear Rate

In a stirred vessel the higher shear rates occur in the trailing vortex pair that exists behind each stirrer blade (Figure 3). From the data given by Van 't Riet and Smith (1975) a maximum shear rate in such a trailing vortex of a Rushton turbine agitator can be determined for both high and low Reynolds numbers (Figure 4). From this figure a maximum shear rate $\dot{\gamma}_{max}$ of about $100N$ can be read at high Reynolds numbers. At the low Reynolds numbers, i.e., in the laminar and transient region of flow, the higher shear rate is about $10N$, which corresponds to the rela-

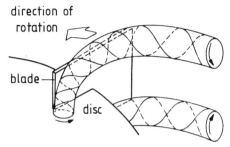

direction of
rotation

blade

disc

Figure 3   Schematic three-dimensional view of the trailing vortex pair in the stirrer blade region (adapted from Van 't Riet and Smith, 1975).

tion of Metzner (Equation 6). It should be noted that Figure 4 is only valid when no gas bubbles are present in the fluid. These will fill up the center of the trailing vortex, which results in a completely different situation. In our insect cell work (Tramper et al., 1986), head space aeration was used; and at the critical stirrer speed air bubbles were not visibly entrained into the medium. The baffles in the vessel prevent vortex formation and thus bubble entrainment. Keeping in mind that a marine impeller and no Rushton stirrer was used, Figure 4 can thus be

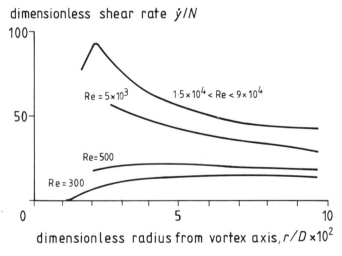

dimensionless shear rate $\dot{y}/N$

$Re = 5 \times 10^3$

$1\cdot5 \times 10^4 < Re < 9 \times 10^4$

$Re = 500$

$Re = 300$

dimensionless radius from vortex axis, $r/D \times 10^2$

Figure 4   The dimensionless shear rate as a function of the dimensionless radius from vortex axis with Re as parameter (adapted from Van 't Riet and Smith, 1975).

used to roughly estimate the maximum shear rate at Re = 1440, which is about 30N. This gives, at the critical stirrer speed N of 9 s^{-1}, a $\dot{\gamma}_{max}$ of 270 s^{-1}.

Croughan et al. (1987) give an equation for the maximum time-averaged shear rate $\dot{\gamma}_{max}$ for the jet off the impeller based on impeller tip speed v_{tip} (m/s):

$$\dot{\gamma}_{max} = c'v_{tip} = c'\pi ND \tag{7}$$

where c' (m^{-1}) is a constant with a value of about 40 m^{-1} for a flat blade impeller. In contrast to the ISF and time-averaged shear rate, cell growth on different scales does not correlate with impeller tip speed. Maximum cell growth or zero cell growth were observed at the same tip speed, depending on the size of the vessel. The impeller tip speed, or this maximum time-averaged shear rate, hence does not appear to play an essential role in this case in determining whether there are detrimental shear effects. However, this does not have to be the case in general. Using Equation (7) for estimation of the maximum time-averaged shear rate at the critical stirrer speed for our insect cell reactor gives $\dot{\gamma}_{max} = 45$ s^{-1}.

In some stirred vessels, however, very high maximum time-averaged shear rates may occur in a region of close clearance between the rotating impeller and a stationary vessel component, e.g., the vessel wall. The tangential flow profile in this region may periodically assume a character similar to the flow between concentric rotating cylinders. The maximum tangential shear rate may then be estimated (Croughan et al., 1989) from the flow profile between concentric rotating cylinders:

$$\dot{\gamma}_{max} = \frac{4\pi N T_v^2}{T_v^2 - D^2} \tag{8}$$

These authors indeed found that the net growth of FS-4 cells on microcarriers was lower in a reactor with such very high maximum time-averaged shear fields. For our insect cell bioreactors a maximum tangential shear rate of 141 s^{-1} is calculated using Equation (8). It should be kept in mind, however, that there is no close clearance and no medium of very high viscosity. Moreover, the baffles also prevent the formation of such flow fields, which further limits the value of this figure.

Average and Maximum Shear Stress

Croughan et al. (1987) show that if one considers the hydrodynamic forces that arise solely from the spatial gradients in (time-averaged) fluid velocity, the maximum shear stress τ_{max} ($N \cdot m^{-2}$) on a microcarrier surface can be estimated by

$$\tau_{max} = 3\eta\dot{\gamma} \tag{9}$$

Similarly, the average shear stress on the surface of a microcarrier can be estimated by (Cherry and Papoutsakis, 1986)

$$\tau_{ave} = 0.5\eta\dot{\gamma} \tag{10}$$

To derive these equations, creeping or Couette flow, i.e., a linear velocity gradient, has to be assumed. If this condition holds for microcarriers, it certainly is true for the much smaller insect cells. If the value 10^{-2} N $\cdot$ s $\cdot$ m^{-2} is substituted for the dynamic viscosity η in Equation (9) and (10), Table 2 can be generated using the shear rates discussed above.

Oh et al. (1989) estimate a maximum turbulent Reynolds shear stress τ_{Re} in the flow close to the Rushton turbine blade from

$$\tau_{Re} = 0.11 \, \rho(\pi ND)^2 \tag{11}$$

The contours of this parameter have been determined by laser Doppler anemometry (Yianneskis et al., 1987) and the values are extremely position and direction/plane dependent. The maximum given by Equation

Table 2 Average and Maximum Shear Stresses at the Critical Stirrer Speed of Insect Cells in Suspension.

Parameter (eqation)	Critical shear rate (s^{-1})	Average shear stress ($N \cdot m^{-2}$)	Maximum shear stress ($N \cdot m^{-2}$)
Integrated shear factor (2)	45	0.23	1.35
Time-averaged shear rate (3)	9.9	0.05	0.3
Maximum shear rate ($\dot{\gamma}_{max} = 30N$)	270	1.35	8.1
Maximum time-averaged shear rate (7,8)	45	0.23	1.35
Maximum tangential shear rate (9)	141	0.71	4.23

(11) occurs in the trailing vortices behind the impeller blade of the Rushton turbine, though much lower values (up to an order of magnitude lower) were found elsewhere, even in the impeller zone. Realizing the limitation that in our insect cell culture a marine impeller instead of a Rushton turbine is used, a value of τ_{Re} of 141 N · m^{-2} is calculated with Equation (11).

A rather accurate quantification of shear stress is possible in a rotaviscometer. Therefore we also used this device to determine the critical shear stress of insect cells in suspension. The viability of suspended insect cells as a function of time at various rotation speeds, thus shear stresses, is given in Figure 5 (Tramper et al., 1986). From this figure it is clear that there is already some loss of viability at 1.5 N · m^{-2} and that this loss considerably increases at higher shear stresses. From Table 2 and τ_{Re} it is clear that in a stirred vessel at the critical stirrer speed an insect cell can in principle experience shear stresses higher than this value of 1.5 N · m^{-2}. However, in contrast to cells in the rotaviscometer, the cells in the stirred vessel are not exposed to these high shear stresses continuously. This may explain why in the stirred vessel the cells are just able to survive at this critical stirrer speed, while in the rotaviscometer at 1.5 N · m^{-2} the killing rate is slightly higher than the growth rate. However, it can be concluded that the critical shear stress at which suspended insect cells start to loose their viability is roughly 1 N · m^{-2}. The value usually mentioned for animal cells on microcarriers is 0.65 N · m^{-2} (Croughan et al., 1987, 1989, and references cited therein), which suggests that they are slightly more fragile than suspended insect cells, which can be expected as argued above. Abu-Reesh and Kargi (1989) mention a value of 5 N · m^{-2}, above which suspended hybridoma cells were damaged in a rotaviscometer in the turbulent regime. This suggests that these cells are somewhat less fragile than our insect cells.

Boundary Layer Shear Stress

Relatively large areas of high shear rate are present in the boundary layers around the solid objects submerged in the stirred vessel. The moving impeller can be expected to have the higher velocity relative to the liquid, and Cherry and Papoutsakis (1986) have analyzed this situation in detail to characterize the general effect of boundary layer shear stress on animal cells adhered to microcarriers. Here, the case of suspended insect cells will be taken as the example. As a first approximation marine and angled flat impeller blades can be modelled as stationary flat plates with fluid moving over them. The Reynolds number for

fraction viable cells

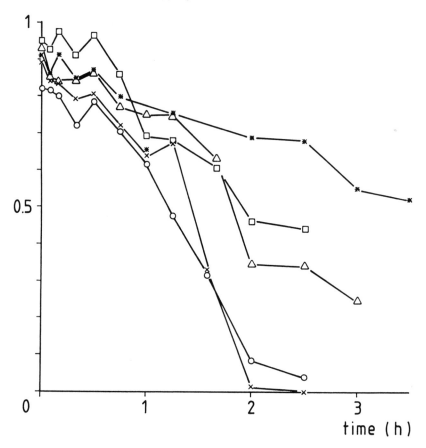

Figure 5 Insect cell viability as a function of time when subject to various shear stresses (N · m^{-2}) in a rotaviscometer: * 1.5; □ 3.4; △ 5.5; × 17; ○ 55 (from Tramper et al., 1986).

transition from laminar to turbulent flow over a flat plate is about 3 × 10^5 (Schlichting, 1979) with

$$\mathrm{Re} = v_\infty L \frac{\rho}{\eta} \tag{12}$$

where v_∞ is the free fluid velocity along the plate (m/s) and L the distance along the plate (m). Cherry and Papoutsakis (1986) take $v_\infty = v_{tip}$.

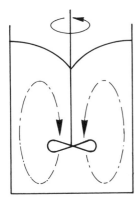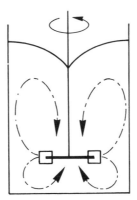

Figure 6   Fluid flow pattern in stirred vessel with marine (left) and Rushton (right) impeller.

If we substitute $v_\infty = v_{tip} = 1.13$ m/s at the critical stirrer speed for insect cells, and $L = 0.02$ m (the width of the impeller blade), a Re of 2260 is calculated, thus a laminar boundary layer can be expected. However, since the flow pattern is as shown in Figure 6 (for comparison the flow pattern of a Rushton stirrer is also shown) for marine and angled flat impeller blades, it is probably better not to take the tip speed, but instead the axial velocity in a screw pump (W. A. Beverloo, personal communication). This axial velocity can easily be estimated by assuming the impellers to be formed like an Archimedean screw. For each distance to the screw axis the same pitch applies. The pitch is the axial displacement of a point that makes on the screw plane a full rotation around the axis. Assume the pitch to be p meters per revolution. If the screw pump operates with N revolutions per second, it causes an axial liquid velocity of about Np meters per second, i.e., at the critical stirrer speed $9 \times 0.12 = 1.08$ m/s (3 blades at 45° angle; effective height 0.04 m; pitch 3×0.04, spiral stairs principle), which accidentally is close to the tip speed in this case. At $L = 0.057$ m (fluid is flowing along the length $(0.04\sqrt{2} = 0.057$ m) of the blades) this liquid velocity gives a Re of 6081, thus also with this value a laminar boundary layer can be expected. However, Schlichting (1979) states that impeller rotation can considerably reduce the Reynolds number for transition. Therefore both laminar and turbulent boundary layers will be considered here like Cherry and Papoutsakis do for the animal cells on microcarriers.

In the case of laminar flow the boundary layer thickness δ(m), defined as the distance from the impeller surface at which the fluid velocity reaches 99% of the free fluid velocity v_∞, is (Schlichting, 1979)

$$\delta = 5 \left(\frac{\eta L}{\rho v_\infty} \right)^{0.5} \tag{13}$$

For the example of insect cells this means:

$$\delta = 0.0152 L^{0.5} \tag{14}$$

The influence of cells is neglected, because Einav and Lee (1973) have shown that 4 and 6 vol % suspensions of neutrally buoyant spheres do not change the boundary layer shape or development from that pre-dicted for "clean" fluids. This certainly is the case then for our insect cell suspensions.

The boundary layer becomes as thick as the diameter of an insect cell (18 μm) at 1.4 μm from the leading edge, Equation (14), which is almost immediately. Cherry and Papoutsakis (1986) arbitrarily assume that the boundary layer must be at least three times the microcarrier diameter in order not to be disrupted by the presence of a bead. For an insect cell of 18 μm this would correspond to a distance L of 12.6 μm from the leading edge. At the trailing edge of the 0.057 m long blade the boundary layer thickness is 3.62 mm, or about 200 cell diameters.

Within the boundary layer the highest shear stress occurs at the solid surface. This laminar wall shear stress τ_{wl} ($N \cdot m^{-2}$) can be calculated by

$$\tau_{wl} = 0.332 \eta v \left(\frac{v_\infty \rho}{\eta L} \right)^{0.5} \tag{15}$$

in which 0.332 is the slope of the dimensionless velocity profile at the wall (Schlichting, 1979). Substituting the appropriate values with $v_\infty = 1.08$ m/s gives

$$\tau_{wl} = \frac{1.178}{L^{0.5}} \tag{16}$$

At the position (12.6 μm) where the boundary layer is three cell diameters thick (54 μm), the shear stress calculated is 102 $N \cdot m^{-2}$, which is far above the critical shear stress. At the trailing edge of the stirrer ($L = 0.057$ m) the wall shear stress calculated is about 5 $N \cdot m^{-2}$, which is still slightly above the critical shear stress. Figure 7 shows that the shear stress rapidly drops to this value of 5 $N \cdot m^{-2}$. From this figure an average thickness of the boundary layer of about 2 mm can be read. The volume

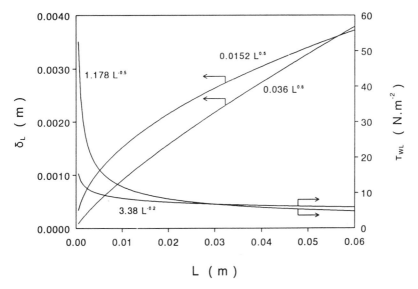

Figure 7 Boundary layer thickness and shear stress on the impeller for laminar and turbulent flow.

over the plate corresponding to this thickness (the distance from the leading to the trailing edge (0.057 m) times and the width of the blades [max 0.02 m] times this thickness) is 2.28×10^{-6} m³. The volume fraction with wall shear stresses slightly above the critical value is thus very small. Furthermore, the time a cell stays in this volume is also very limited. This, together with the flexibility of shape of the cells which helps to accommodate short beatings, can explain how the cells survive these wall shear stresses.

For turbulent boundary layers the formulas for boundary layer thickness δ(m) and wall shear stress τ_{wl} (N · m^{-2}) are (Schlichting, 1979):

$$\delta = 0.37L \left(\frac{\rho v_x L}{\eta} \right)^{-0.2} \tag{17}$$

and

$$\tau_{wl} = 0.0294 \, \rho v_x^2 \left(\frac{\rho v_x L}{\eta} \right)^{-0.2} \tag{18}$$

Substituting the appropriate values for insect cells with $v_x = 1.08$ m/s gives

$$\delta = 0.036 L^{0.8} \tag{19}$$

and

$$\tau_{wl} = \frac{3.38}{L^{0.2}} \tag{20}$$

In Figure 7 these functions are compared to their laminar equivalents. The turbulent shear stresses are considerably lower than the laminar ones over the first half of the impeller blade. However, like the laminar ones, the turbulent wall shear stresses exceed the critical value of $1 \text{ N} \cdot \text{m}^{-2}$ at all values of L. One could thus draw the same conclusion about the damaging effect of these turbulent wall shear stresses as above for the laminar wall shear stresses, i.e., that it can be neglected.

The shear field in the boundary layer also causes the cell to rotate. Assuming Couette flow, Equations (10) and (11) can be used to estimate the maximum and average shear stress, respectively, on the rotating cell. For a linear velocity gradient of 1.08 m/s over a boundary layer of about 2 mm (average thickness for the example, see Figure 7), $\dot{\gamma} = 1.08/(2 \times 10^{-3}) = 540 \text{ s}^{-1}$. The average and maximum shear stress thus calculated for the insect cell culture medium ($\eta = 10^{-2} \text{ N} \cdot \text{s} \cdot \text{m}^{-2}$) with Equations (10) and (11) are 3.6 and $21.6 \text{ N} \cdot \text{m}^{-2}$. Like the wall shear stresses these are above the critical value and it is therefore also unlikely that the shear stresses as result of rotation are involved in the breaking of the cells to any considerable extent.

Smallest Turbulent Eddy Model

Introduction

The effects of hydrodynamic forces on animal cells in a stirred bioreactor have been most extensively studied in microcarrier systems, where the cells grow attached to the surface of spherical particles typically about 180 μm in diameter. The mechanisms that best explain the experimental results are interactions of the microcarrier beads with turbulent eddies having length scales smaller than the beads, and collisions between the beads (Croughan et al., 1987; Cherry and Papoutsakis, 1988; Croughan et al., 1989). If these same explanations of damage mechanisms that work well for cells on microcarriers are applied to fragile cells in suspension, they fail (Cherry and Kwon, 1990). Individual cells of about 15 μm, which is a diameter typical for animal cells, are much smaller than the length scale of the smallest turbulent eddy in any reasonably agitated bioreactor (Croughan et al., 1987; Cherry and Papoutsakis, 1988; Oh et al., 1989) and therefore would be expected to be insensitive to damage from the eddy interactions that damage cells on microcarriers. Similarly, cells at a density of 10^{12} cells $\cdot$ m^{-3}, which is typical for animal cell cultures when no perfusion is applied, occupy only about 0.1 vol % of the bioreactor and should have a negligible number of collisions, especially since their density is so close to that of the medium that they will not deviate greatly from the fluid stream lines. However, cells in suspension can also be susceptible to excessive agitation, although the general level of sensitivity seems to be less than for anchored cells (Cherry and Kwon, 1990). In particular, when the serum concentration is lowered in the culture medium, which is often done for economic reasons and for ease of downstream processing, the shear sensitivity increases and agitation can rapidly become too high if no other protective agents are added. It is thus a situation to be reckoned with, and therefore the pertinent analysis of Cherry and Kwon (1990) is summarized here, again taking insect cells as a reference. In their article they present an analysis of the time-varying shear stresses imposed on a spherical cell entrained in a model turbulent eddy much larger than the cell.

The Eddy Model

The common physical picture of turbulence starts with large eddies created, in the case of a mechanically stirred vessel, by an impeller (Figure 3). These large eddies pass their kinetic energy on to successively smaller eddies without loss until the energy is finally dissipated viscously as heat in eddies of some smallest size. In case of isotropic turbulence, which has no preferred direction, these smallest eddies have characteristic scales of length λ_K and velocity v_K (Kolmogorov theory):

$$\lambda_K = \left(\frac{v^3}{\epsilon} \right)^{0.25} \tag{21}$$

$$v_K = (\epsilon v)^{0.25} \tag{22}$$

with ϵ the empirical mass average of turbulent energy dissipation (m$^2 \cdot$ s^{-3}) and v the kinematic viscosity (m$^2 \cdot$ s^{-1}). The physical structure of these eddies can in essence be pictured as long, flexible, rotating cylinders of fluid which are stretched along their axis by larger scale flows. This axial stretching causes a radially inward flow of the fluid in and around the eddy (see Figure 9). A steady state flow field with these properties, which also satisfies the Navier-Stokes and continuity equations, is the Burgers vortex described by Cherry and Kwon (1990):

$$v_r = -0.5\alpha r \tag{23}$$

$$v_z = \alpha z \tag{24}$$

$$v_\theta = \frac{2v\omega_0}{\alpha r} \left[1 - \exp \left(\frac{-\alpha r^2}{4v} \right) \right] \tag{25}$$

with α a strain rate parameter in the axial direction (s^{-1}) and ω_0 a vorticity parameter (s^{-1}). To match α to actual turbulence, Cherry and Kwon (1990) use for α the root mean square value of the strain rate in isotropic turbulence. Based on experimental data α has been suggested to be

$$\alpha = 0.18 \left(\frac{\epsilon}{v} \right)^{0.5} \tag{26}$$

for stirred laboratory vessels, while the exact value is $0.26(\epsilon/v)^{0.5}$ (Cherry and Kwon, 1990). By equating the dissipative work of the model flow field with that measured in an experimental system, and by assuming that the flow field of Equations (23–25) completely fills the volume around each eddy, Cherry and Kwon (1990) derive for the vorticity parameter

$$\omega_0 = 0.95I^{-1} \left(\frac{\epsilon}{v} \right)^{0.5} \tag{27}$$

and for the local tangential shear rate

$$\dot{\gamma} = \omega_0 \left[-\frac{1}{R} + \left(\frac{1}{R} + 1 \right) exp(-R) \right] \tag{28}$$

A measure of the contribution of shear to the average dissipation throughout the whole eddy can be described by

$$I = \left\{ \frac{1}{L'} \int_0^{L'} \left[-\frac{1}{R} + \left(\frac{1}{R} + 1 \right) exp(-R) \right]^2 dR \right\}^{0.5} \tag{29}$$

R is the dimensionless radius defined as

$$R = \frac{\alpha r^2}{4v} \tag{30}$$

To evaluate the expression for I, the distance L' (in the dimensionless units of R) to the edges of the coherent flow field must be known. In Figure 8 the dimensionless tangential velocity, Equation (25), and shear rate, Equation (28), are plotted as a function of R. The maximum tangential velocity in the eddy occurs at $R \approx 1.27$ and the maximum shear rate at $R \approx 1.79$, so the edge of the eddy is expected to be at values of R significantly greater than these. The actual extent of the eddy might be estimated from the observation that the maximum contribution to a Fourier analysis of turbulent velocity gradients occurs at a wavelength of about $40(\approx v^3/\epsilon)^{0.25}$ (Cherry and Kwon, 1990). If this is taken to be the

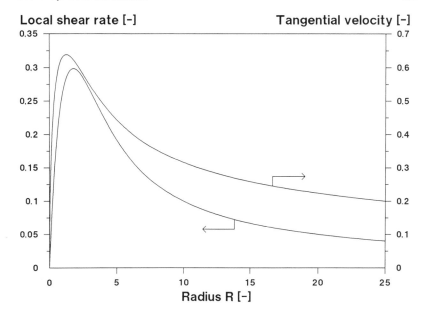

Figure 8 Dimensionless tangentional velocity $(v_\theta/\omega_o)(\alpha/\nu)^{0.5}$ and shear rate $(\dot{\gamma}/\omega_o)$ in the model eddy as a function of the dimensionless radius $R = \alpha r^2/4\nu$ (adapted from Cherry and Kwon, 1990).

typical distance between the centers of two nearby eddies, since the independent high velocities near the eddy cores would cause the highest relative velocities and hence gradients, the extent of the average eddy can be estimated by $20(\nu^3/\epsilon)^{0.25}$, or $L' = 26$. This half spacing of 20 Kolmogorov length scales is well in line with a theoretical analysis of turbulence structure indicating that eddies $12(\nu^3/\epsilon)^{0.25}$ in diameter are the most active in energy dissipation, and with experimental measurements suggesting that the smallest eddies have actual diameters 10–15 times the Kolmogorov characteristic length scale (Cherry and Kwon, 1990). At this presumed value of $L' = 26$ for the eddy extent, $I = 0.133$, from Equation 29, which leads to

$$\omega_0 = 7.14 \left(\frac{\eta}{\nu} \right)^{0.5} \tag{31}$$

We are ultimately interested in following a small cell as it moves through this eddy flow field to learn the forces to which it is subjected, with the local tangential shear rate, Equation (28), as the focus of attention. This shear rate is a function of α, ω_0 and r. Values for ϵ and v are presumed to be known from the particular experimental conditions. The cell's radial position is found by solving Equation (23), which is independent of the other two equations for velocities in the eddy

$$r = r_0 exp(-0.5\alpha t) \tag{32}$$

The constant r_o is an arbitrary starting point for the cell's position at time zero.

Cell-Eddy Interactions

In their analysis Cherry and Kwon (1990) assume that the cell remains spherical during its interaction with the eddy's internal flow field and that the disturbance of this flow field by the cell is insignificant. Furthermore, they assume Stokes flow as the Reynolds number

$$Re = \frac{\rho \dot{\gamma} d_c^2}{\eta} \tag{33}$$

has a value of about 0.01 at a shear rate $\dot{\gamma}$ of $100\ s^{-1}$ and a cell diameter of 15 μm, which is typical of the range considered. For our insect cell suspension ($d_c = 18\ \mu m$; $\rho = 1000\ kg \cdot m^{-3}$; $\eta = 0.01\ N \cdot s \cdot m^{-2}$) Re is even smaller at $\dot{\gamma} = 100\ s^{-1}$, i.e., about 0.003. The flow field around a sphere in such a linear (Couette) shear field with shear rate can be described analytically. In their derivation of shear stresses, Cherry and Kwon (1990) use linear rather than angular velocities with coordinates as pictured in Figure 9, which shows the trajectory of a cell in an eddy

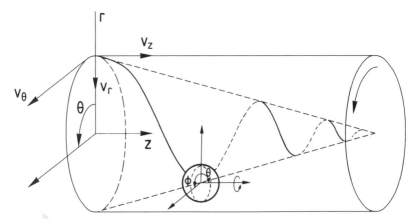

Figure 9   Trajectory and coordinate systems of a cell in the model eddy.

and the axis around which the cell rotates. Using generalized expressions for shear stresses in spherical coordinates and solving for the stresses on the sphere surface, their most important result for the magnitude of the shear stress is

$$\tau = 2.5\eta\ \dot{\gamma}\ \sin\theta[1 - \sin^2(\dot{\gamma}t)\sin^2\theta]^{0.5} \tag{34}$$

with $\dot{\gamma}$ being specified by Equations (26), (27), (28) and (32). The only physical variables that must be specified to calculate this shear stress are μ, ϵ, ν and Θ. The value of r_o in Equation (32) is arbitrary and effects only the value of t at which the shear stress begins to increase, not the intensity or duration of the event. At the equator of the rotating sphere where the shear stress is maximal, $\Theta = 90°$, and Equation (34) reduces there to the simple form

$$\tau = 2.5\eta\dot{\gamma}|\cos(\dot{\gamma}t)| \tag{35}$$

The normalized values of the shear stress, (Equation 34), at points on the sphere's surface during its interaction with an eddy are shown in Figure 10. Near the poles of the rotational axis (Θ small) the magnitude of the shear stress shows a minimal cycling about a baseline value that increases with distance from the pole. Figure 11 shows the temporal

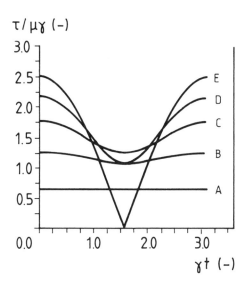

Figure 10 Dimensionless shear $\tau/(\mu\dot{\gamma})$ as a function of θ (in degrees) and dimensionless time $\dot{\gamma}t$ (adapted from Cherry and Kwon, 1990). Lines A–E are 15, 30, 45, 60 and 90 degrees, respectively.

shear stress (Nm^{-2})

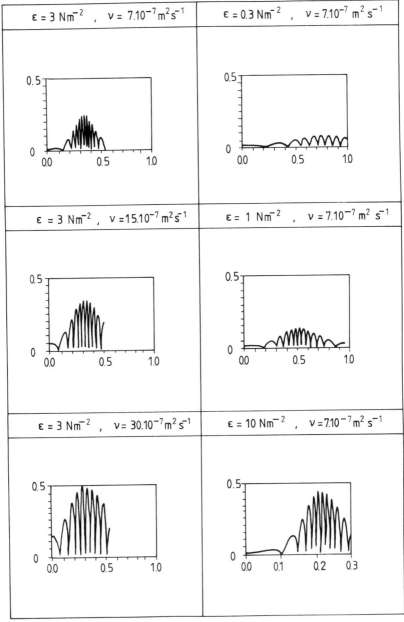

Figure 11 Time-varying shear stress at a point on the equator of the rotating sphere (adapted from Cherry and Kwon, 1990).

behavior of the shear stress along the equator of the sphere with the major bioreactor variables ϵ and ν as parameters. At early times in the interaction the local shear rate is low, and the shear stress at a point cycles slowly and with a low magnitude. As the cell moves closer to the center of the eddy (Figure 9), the shear rate increases (Figure 8), and both the magnitude and the frequency of the shear stress increase too. Eventually the cell moves into the region of solid body rotation at the core of the eddy, and the local shear rate and the shear stress both decrease to near zero. Higher values of ϵ cause a shorter duration event with higher maximum shear stresses and faster cycling (Figure 11, right column). Raising the kinematic viscosity ν leads to a longer event with a slower cycling but with a higher maximum shear stress in each peak (Figure 11, left column). This variation of the shear stress with these major bioreactor variables can be described quantitatively. The maximum shear rate occurs at $R = 1.79$ (Figure 8). Substituting this into Equation (28) to find $\dot{\gamma}_{max}$ and taking $\cos(\dot{\gamma}t) = 1$, Equation (35) gives the maximum shear stress as

$$\tau_{max} = 5.33\rho(\epsilon\nu)^{0.5} \tag{36}$$

The maximum frequency f of the shear stress comes from the argument of the $\cos t$ $(\dot{\gamma}t)$ term in the same equation, accounting for the two peaks per sinusoidal cycle caused by the absolute value:

$$f = 2\left(\frac{\dot{\gamma}_{max}}{2\pi}\right) = 0.678\left(\frac{\eta}{\nu}\right)^{0.5} \tag{37}$$

In addition to this high frequency shear stress variation caused by the cell's rotation in the eddy flow field, there is a more irregular, lower frequency component, resulting from the cell's passing from one eddy to the next. The time it takes a cell to circulate around the bioreactor, from the high turbulence zone around the impeller through a more quiescent area near the liquid surface and back to the impeller, would add another even lower frequency peak in the power spectrum, depending very much on the specific bioreactor considered. Cherry and Kwon (1990) state that they have not made attempts to model the latter two, but signal the need for investigation of the response of cells to shear stress stimuli of a range of frequencies. They also stress that, even though the eddy model they use in their analysis is a standard one for turbulence and mixing studies because of its simplicity and generally appropriate behavior, their model has not been experimentally verified and there is no essential reason that the individual eddies must be at steady state. Nevertheless, they conclude that the use of this model should be acceptable for estimating the range of stresses on a suspended cell in turbulence. We agree with that, moreover, because it greatly enhances the insight in these complex phenomena.

Application to a Suspended Insect Cell

For application of the above theory of Cherry and Kwon (1990) to our insect cell reactors, we have to estimate the energy dissipation in the vessel, which is equal to the power consumption P_s(W). The general equation for the power consumption is

$$P_s = N_p \rho N^3 D^5 \tag{38}$$

For fully turbulent conditions (Re $\geq$ 10000), the dimensionless power number N_p for any stirrer type in a baffled vessel is constant (Van 't Riet and Tramper, 1991). For lower Re numbers N_p is a function of Re only. At the critical stirrer speed ($N = 9$ s^{-1}) Re calculated by Equation (4) is 1440, thus far below fully turbulent conditions. From the appropriate graph in the above reference an N_p of about 1 can be read at this Re number. Substituting this in Equation (38), together with $\rho = 1000$ kg $\cdot$ m^{-3}, $N = 9$ s^{-1}, and $D = 0.04$ m gives a P_s of 2.3×10^{-3} W, which in a stirred vessel of 1 dm^3 and a density of 1000 kg $\cdot$ m^{-3} is equal to the mass average rate of energy dissipation ϵ (W $\cdot$ kg^{-1}). With a kinematic viscosity v of 10^{-5} m$^2 \cdot$ s^{-1} this yields for the mean Kolmogorov length scale, via Equation (21), $\lambda_K = 1.44$ mm, which is two orders of magnitude larger than an insect cell. The energy dissipation near the impeller is however much bigger. Oh et al. (1989) assume, as others have, that essentially all the energy is dissipated in half the volume occupied by the impeller. In that case, the Kolmogoroff length scale in that region of the impeller is given by

$$\lambda'_K = \left(\frac{v^3}{\epsilon'} \right)^{0.25} \tag{39}$$

where

$$\epsilon' = 130\epsilon \tag{40}$$

which gives a minimum Kolmogoroff length scale, via Equation (34), of $\lambda = 0.43$ mm, still considerably larger than an insect cell. This thus does not conflict with the assumptions of Cherry and Kwon. Calculation of the mean maximum shear stress by means of Equation (36) yields 2.5 N $\cdot$ m^{-2}, which is very close to the critical shear stress for insect cells. Multiplying the mean energy dissipation first by a factor 130, Equation (40), yields for the maximum shear stress 29 N $\cdot$ m^{-2}, which clearly is significantly above the critical value. From this, one could conclude that the maximum shear stress obtained from the mean energy dissipation is the more likely parameter for scale-up than the one obtained from the maximum energy dissipation in the impeller region, and therefore earns further attention.

To calculate turbulent shear stress, Oh et al. (1989) use

$$\tau = \rho v_\lambda^2 \tag{41}$$

This equation was derived by Levich (1962), also based on Kolmogoroff's theory, with v_λ the fluctuating eddy velocity for an eddy of size λ. For an eddy size λ greater than λ_K

$$v_\lambda = (\epsilon\lambda)^{0.33} \tag{42}$$

and for an eddy of size λ less than λ_K

$$v_\lambda = \left(\frac{\epsilon}{v}\right)^{0.5} \lambda \tag{43}$$

When taking a stress equal to that of an eddy of size λ_K, i.e., from Equation (21) and either Equation (42) or (43) via Equation (41), this gives

$$\tau_{\lambda_K} = \rho(\epsilon v)^{0.5} \tag{44}$$

Substituting the appropriate values of our insect cell suspensions at the critical stirrer speed gives $0.15 \text{ N} \cdot \text{m}^{-2}$. This is rather close to the critical shear stress and this parameter is thus worth further investigation too.

Oh et al. (1989) argue that in general, for such topics as liquid-liquid dispersion, solid suspensions and gas dispersion in stirred vessels, it is usual to consider that the eddy size of importance for a particular process is related to the size of the dispersed phase, i.e., in this case the size of an insect or animal cell in general, d_c, so that from Equations (41) and (43) it follows

$$\tau = \rho \left(\frac{\epsilon}{v}\right) d_c^2 \tag{45}$$

This equation holds except in the unlikely case that at the highest local energy dissipation rate, via Equation (40), $d_c > \lambda_{K'}$. Equations (41) and (42) then give

$$\tau = \rho(\epsilon' d_c)^{0.67} \tag{46}$$

The diameter of an insect cell is about 18 μm, thus considerably smaller than λ_K and even $\lambda_{K'}$ at the critical stirrer speed (see above). Substituting in Equation (45) other values for insect cell suspensions gives a very low shear stress of 7.5×10^{-5} N · m^{-2}. Equation (46), on the other hand, yields 0.3 N · m^{-2}, which in contrast is close to the critical value. Keeping in mind the above limitation, some further attention to this parameter is recommended.

Implications for Reactor Design

In the introduction to this section the reasons for stirring a bioreactor are given. Firstly, it is required to prevent settling of the cells. As a rule of thumb one can say that the minimum velocity in the bulk phase should be at least twice the terminal settling velocity of the cell. Stokes' law can be used to calculate the settling velocity v_s (m/s) of a single cell as:

$$v_s = \frac{d_c^2(\rho_c - \rho)g}{18\eta} \tag{47}$$

in which ρ_c is the specific density of the cell (kg · m^{-3}). In order for Stokes' law to be valid, the cell Reynolds number, defined as

$$\text{Re} = \frac{\rho v_s d_c}{\eta} \tag{48}$$

should be less than 1.

In the intermediate range ($1 < \text{Re} < 10^3$), Beek and Mutzall (1975) give as relation for v_s

$$c_w \text{Re}^2 = \frac{4d_c^3\rho(\rho_c - \rho)g}{3\eta^2} \tag{49}$$

in which c_w is the drag coefficient ($-$). The right hand side of Equation (49) can be calculated for a given situation, and by means of Figure 12 the relevant Re can be found, from which v_s can be calculated. Due to the very small difference in density of cell and medium, keeping cells in suspension generally requires very gentle stirring. For our insect cell suspensions, assuming a density difference of 25 kg · m^{-3}, a v_s of 4.4×10^{-7} m/s is calculated via Equation (47), and with that a Re of 8×10^{-7}, via Equation (48). Thus, Equation (47) is valid for this situation. Extremely low fluid velocities are thus required to keep the insect cells in suspension, and this should not create problems from the point of view of shear sensitivity. Even cells attached to microcarriers generally need only very gentle stirring to keep them in suspension.

According to Cherry and Papoutsakis (1986), the primary reason for stirring cell culture reactors is transfer of oxygen and maintaining homo-

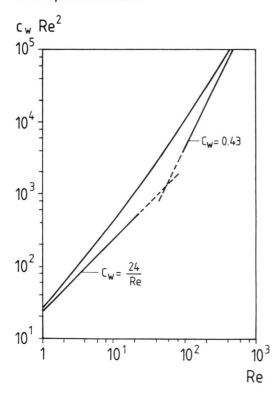

Figure 12 Drag coefficients and related functions for spherical particles (adapted from Beek and Mutzall, 1975).

geneity by minimizing variations throughout the reactor of dissolved oxygen and other nutrient concentrations or temperature. The average liquid velocity needed to give effective homogeneity can be estimated by requiring that the cell moves through the various areas of different conditions in an amount of time that is small compared to their biological response time. Cherry and Papoutsakis (1986) refer to papers concerning microbial cells from which a minimum response time of 2–3 seconds can be derived. Assuming that a 1 dm³ bioreactor has a characteristic dimension of about 0.1 m, the minimum liquid velocity needed to ensure apparent homogeneity is then of the order of 0.05 m/s, which generally means that the bulk liquid flow is turbulent. As the scale increases, liquid velocities, and thus turbulence, should be higher to ensure homogeneity. An analysis of shear and turbulence as given in this

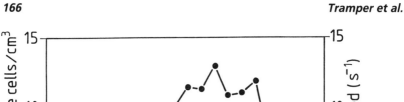

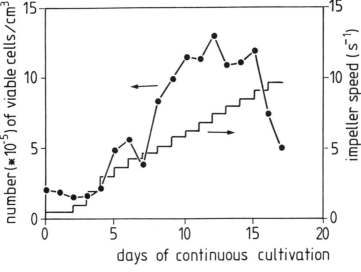

Figure 13 Insect cell density in a continuous culture at various stirrer speeds (adapted from Tramper et al., 1986).

chapter is thus generally essential for rational design and scale-up of bioreactors for growth of fragile cells.

Stirring is also required to enhance oxygen supply. Significant improvement can be accomplished by dispersion of the air bubbles via sparging, which is unavoidable if the size of the bioreactor increases. As a rule of thumb, a tip speed of the impeller of about 2 m/s is needed. Figure 13 shows that insect cells die in a small stirred bioreactor (1.5 dm^3) if the impeller speed is larger than about 9 s^{-1}, which means a tip speed of 1.13 m/s ($D = 0.04$ m). We can therefore conclude that stirring, as a means of substantially improving oxygen transfer by dispersion of air bubbles in insect cell suspensions, but also for animal cell suspensions in general, is usually impossible. Therefore we have directed our research mainly to bubble column and airlift bioreactors.

6.3 BUBBLE COLUMN AND AIRLIFT BIOREACTORS

Introduction

In contrast to the impeller stirred vessels, bubble column and airlift bioreactors have been used relatively rarely in determining and quantifying the fragility of cells, even though they are of great interest for growth of fragile cells (Katinger and Scheirer, 1982). In fact, only three or four groups have studied and used bubble columns and airlift bioreactors in this respect. Handa-Corrigan et al. (1989) studied the effects of sparging on hybridomas and other mammalian cells in suspension culture. They found that damage of cells occurs especially during the bursting of the air bubbles at the suspension surface, and that the nonionic surfactant Pluronic has a concentration-dependent protective effect. The latter phenomenon has been studied extensively and many recent papers describe and quantify this effect of Pluronic and other polymers for insect cells (e.g., Goldblum et al., 1990) and animal cells in general (Papoutsakis, 1991). Also, the protective effect of serum or protein against agitation and aeration is the subject of many recent papers, e.g., (Hülscher et al., 1990; Van der Pol et al., 1990; 1991). These protective effects will not be dealt further with in this chapter.

Wudtke and Schügerl (1987) investigated the fragility of insect cells using various methods, including a bubble column. In agreement with the findings of Handa-Corrigan et al. (1989), these authors found that covering the suspension with a paraffin layer prevented the appearance of cell debris, indicating that the bubble bursting is indeed the damaging process. Quantitative relationships suitable for bioreactor design and scale-up are not given by these two groups. Good growth on a larger scale is however possible, both for insect cells (Maiorella et al., 1988) and animal cells in general (Birch and Arathoon, 1990).

Estimations of Shear

Local Shear

As stated at the end of the previous section, our incentive to study bubble column and airlift bioreactors for growth of insect and other fragile cells stems from the observation that stirring from the point of view of maximum oxygen transfer is impossible, while supply of oxygen is the crucial step at larger scales. In our early papers on this topic (Tramper and Vlak, 1988; Tramper et al., 1986) it is stated that we repeatedly failed to grow insect cells in a small airlift bioreactor (0.5 dm^3), even though the estimated maximum shear stress (0.6 N · m^{-2}) associated with fluid flow was slightly below the shear stress critical for growth of the cells. This maximum shear stress was estimated using the equation:

$$\tau = 0.5 \, \rho v^2 k_w \qquad (50)$$

The maximum fluid velocity v was measured to be 0.03 m/s, and the resistance coefficient k_w was taken as 1.3 for a sharp bend where it is highest (Beek and Mutzall, 1975). From this it was concluded that the injection, the rising, and/or the bursting of the air bubbles was responsible for the faster death rate than growth rate. To get a rough idea of what shear stresses an insect cell could have experienced as result of these three processes, Equation (1) can be used. For that it is arbitrarily assumed that on one side of the insect cell the fluid velocity is equal to that of a nearby air bubble, and that on the other side it is zero. At the injector the velocity v_{bi} (m/s) of the air bubbles can be calculated from the volumetric airflow rate F_g (m³/s) and the inner diameter d_i of the injector nozzle:

$$v_{bi} = \frac{F_g}{0.25\pi d_i^2} \tag{51}$$

With a rather low airflow of 1 dm³/h and a nozzle diameter of 1 mm, v_{bi} is calculated to be 0.35 m/s. Substituting this for dv in Equation (1), the insect cell diameter, i.e., 18 μm, for dx, and 0.01 N · s · m^{-2} for η, the shear stress thus calculated is 196 N · m^{-2}. This is orders of magnitude above the critical shear stress level at which insect cells start to die off.

The velocity of a rising air bubble is of the order of 0.25 m/s (Heijnen and Van 't Riet, 1984). A calculation similar to the one above yields a shear stress of about 139 N · m^{-2}, which again is far above the critical level.

The situation of the bursting air bubble at the surface of the cell suspension is more complex. The air bubble forms a "hill" under the liquid surface (Figure 14). As soon as the surface tension of the liquid

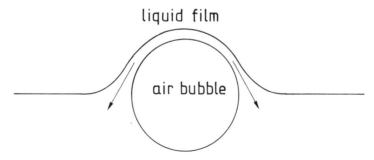

Figure 14 Bursting of an air bubble.

film above the air bubble becomes too small, this film breaks and the fluid flows back to the bulk. The associated fluid velocity v_{bb} (m/s) can be estimated by the Culick equation (Havenbergh and Joos, 1983):

$$v_{bb} = 2 \left(\frac{\sigma}{\rho d_f} \right)^{0.5} \tag{52}$$

The surface tension σ (N/m) and the thickness d_f (m) of the liquid film can be assumed to be on the order of 0.035 N $\cdot$ m^{-1} and 40 μm, respectively. Equation (52) thus gives 1.87 m $\cdot$ s^{-1} for the liquid velocity v, which yields, with Equation (1), a shear stress of 1039 N $\cdot$ m^{-2}. That the shear stress estimated for the bursting process is the highest corresponds to the above-mentioned findings that cell killing occurs at the surface of an aerated suspension.

The above estimates of the shear stresses that an insect cell could in theory experience when sparging the suspension with gas bubbles, and arbitrarily assuming that on one side of the cell the fluid velocity is zero and on the other side equal to that of a nearby gas bubble, clearly shows that the estimated shear stress is far above the critical value. The observation that not all cells are instantly broken can be explained by the fact that either the arbitrary assumption is far off reality, or that the critical value of 1 N $\cdot$ m^{-2} applies to the situation in a rotaviscometer where the cells are constantly exposed to this shear. In the case of sparging, high shear forces, possibly in the range of the estimate above, are only momentarily affecting the cells. Apparently the flexibility in shape of the cells is enough to accommodate most of these short "beatings," or the frequency of the blows is too low. Obviously, another explanation is that the killing process as result of sparging the suspension is based on a different mechanism.

Average Shear Rate

A commonly used expression (Nishikawa, 1991) for the average shear rate in a bubble column reactor is

$$\dot{\gamma}_{ave} = 5000 u_g \tag{53}$$

where u_g is the superficial gas velocity (m $\cdot$ s^{-1}). This equation has also been applied to airlift reactors (Popovic and Robinson, 1987; 1988) by replacing u_g with the superficial gas velocity in the riser u_{gr} (m $\cdot$ s^{-1}). Chisti and Moo-Young (1989) state in their reaction to the papers of Popovic and Robinson that the use of Equation (53) for bubble columns remains questionable, but that its extension to airlift devices is quite inappropriate. However, a recent communication (Allen and Robinson, 1991) refutes the major criticism. As there is not really a better and easier way to estimate average shear rates, we will use it here to at least roughly estimate the magnitude of shear in our bubble column and airlift studies. Most of our shear sensitivity studies with insect cells have been in small bubble columns. In a first series of experiments the airflow rate (Tramper et al., 1986) was varied (1, 3, 5 and 7 dm³/h) in a bubble column (height 0.18 m; inside diameter 0.036 m) containing an insect cell suspension. Using Equations (53) and (1) ($\eta = 0.01$ N · s · m^{-2}) gives the shear stress for the 4 airflows: 0.0144, 0.0433, 0.0722, and 0.1011 N · m^{-2}, respectively. Even at the lowest airflow a slight loss of viability was observed, although the estimated shear stress is significantly less than the critical value of 1 N · m^{-2}. In another series (Tramper et al., 1988) the diameter of the bubble column was varied (0.035, 0.052, 0.062 and 0.08 m) keeping the height and airflow (7 dm³ · h^{-1}) constant. Again using Equations (53) and (1) gives 0.0955, 0.0458, 0.0322 and 0.0193 N · m^{-2}, respectively, and again a slight loss of viability was already seen at the lowest shear stress. In a third series (Tramper et al., 1987) the height of the bubble column was varied, keeping all the other parameters constant. Thus, there were no changes in the superficial gas velocity, or the shear stress estimated this way (0.0683 N · m^{-2}). However, clear differences in death rates were observed, proving that shear stress estimated this way is not a good scale-up parameter.

Assuming the gas expansion is the main contributor to power input for a bubble column, the power per unit volume, P_g/V, is

$$\frac{P_g}{V} = \rho g u_g \tag{54}$$

which is then also equal to the empirical average of turbulent energy dissipation ϵ. Similar to the shear stress estimated by Equation (53), this is not a good scale-up parameter, as both depend on the superficial gas velocity only (ρ and g are constant). The same applies to airlift reactors.

Killing-Volume Hypothesis

In order to find correlations between bubble column design parameters and death rate of the cells, we have derived a model on the basis of the following two assumptions: firstly, that the loss of viability of the cells due to aeration is a first order process:

$$C_x(t) = C_x(0)e^{-k_d t} \tag{55}$$

in which C_x is the concentration of viable cells (number, kg or mol per m^3) and k_d the first order death rate constant (s^{-1}). Secondly, we assume that associated with each air bubble, during its entire life time, is a hypothetical volume V_k(m^3) in which all viable cells are killed:

$$-V \frac{dC_x}{dt} = nC_xV_k = \frac{F_g}{(\frac{1}{6})\pi d_b^3} C_xV_k \tag{56}$$

with n the number of air bubbles generated per second (s^{-1}) and d_b the diameter of the air bubbles (m). Separation of variables and integration, assuming V_k to be constant in time, yields:

$$C_x(t) = C_x(0)exp\left(\frac{6F_gV_k}{\pi d_b^3 V}t\right) \tag{57}$$

Combining Equations (55) and (57) gives:

$$k_d = \frac{6F_gV_k}{\pi d_b^3 V} \tag{58}$$

or

$$k_d = \frac{24F_gV_k}{\pi^2 d_b^3 T_v^2 H} \tag{59}$$

with H the height of the liquid in the bubble column (m). In order for these equations to become useful, the dependence of k_d and V_k on F_g, d_b, T_v and H was experimentally determined (Tramper et al., 1986, 1987 and 1988). It was found that k_d is proportional to F_g, T_v^{-2} and H^{-1}. This implies that V_k must be independent of these variables. That V_k is independent of H means that the shear stresses associated with the rising of the air bubbles can be assumed to have a negligible damaging effect, a conclusion also reached for hybridoma, myeloma and baby hamster kidney cells (Emery et al., 1987). The estimations of the various shear stresses already revealed that shear associated with the rising of air bubbles is relatively low in comparison to that associated with the injection of air bubbles and especially their bursting at the surface. Although experimentally much more difficult to establish, it has also been found (Tramper et al., 1988) that k_d is largely independent of the air bubble diameter. This means that V_k must be proportional to d_b^3 (or the volume of the air bubble). By introducing a specific hypothetical killing volume

$$V_k' = \frac{V_k}{(\frac{1}{6})\pi d_b^3} \tag{60}$$

which is the hypothetical killing volume divided by the volume of one air bubble, Equation (59) can be reduced to

$$k_d = \frac{F_g V'_k}{(\frac{1}{4})\pi T_v^2 H} = \frac{F_g V'_k}{V} \tag{61}$$

If V_k or V'_k is known, and they can be measured very simply in a small bubble column (Tramper et al., 1988), k_d can be easily calculated for each desired bubble column by Equation (61). The average calculated hypothetical killing volume V_k is 4.6×10^{-10} m^3 for our insect cells, which corresponds to a spherical volume with a diameter of about 1 mm.

It is clear that the killing volume hypothesis has no physical basis. However, at the 1990 AIChE meeting in Chicago, D. Orton from the group of D.I.C. Wang (MIT) proposed a mechanism of bubble bursting. In this process of bursting, after liquid-film breakup (Figure 14), a liquid droplet of about 1–2 mm is carried off into the air. The size of these droplets is about the same as the killing volume. However, the assumption that all cells in the ejected droplet are killed has not yet been proven. At the same meeting, J. J. Chalmers from Ohio State University presented the work of his group on the microscopic visualization of insect cell/air bubble interactions (Bavarian et al., 1991; Chalmers and Bavarian, 1991). What they observed is that an accumulation of cells occurred in the wake of the air bubbles, which is the liquid most likely to be shot off into the air during the bursting process, according to the mechanism presented by Orton. The latter mechanism is different from earlier work by MacIntyre (1972), who showed that the liquid in the upward jet developing after film breakup originates from the film and the rim of the bubble. In theory the findings of MacIntyre, Orton, and Chalmers could give a physical basis to our killing volume model. Obviously more research is needed to prove it. What we have shown, however, is that this model is also valid to several hybridoma cell lines and to airlift reactors (Martens et al., 1990 and 1991; Van der Pol et al., 1990 and 1991; Jöbses et al., 1991). Experimental validations in larger bubble column and airlift bioreactors still remain to be done.

Implications for Reactor Design

For scale-up of fragile cell cultures in which oxygen is supplied by sparging air through the suspension, it is important to correlate shear sensitivity with the oxygen need of the cells. The specific oxygen transfer rate OTR (mol $\cdot$ m^{-3} $\cdot$ s^{-1}) can be written as

$$\text{OTR} = k_{ol}A(C_o^* - C_o) = \text{OUR } C_x \tag{62}$$

with k_{ol} the oxygen transfer coefficient (m/s), A the specific surface area of the air bubbles ($m^2 \cdot m^{-3}$), C_o^* the concentration of oxygen in the liquid when in equilibrium with air, C_o the actual concentration (mol $\cdot$ m^{-3}), and the oxygen uptake rate (OUR) of one unit of cells (mol oxygen per unit number, mol, or kg of cells). A can also be written as

$$A = \frac{n' \pi d_b^2}{(\frac{1}{4}) \pi T_v^2 H} \tag{63}$$

with n' being the number of air bubbles present in the reactor:

$$n' = \frac{F_g H}{(\frac{1}{6}) \pi d_b^3 v_r} \tag{64}$$

where v_r is the rising velocity of an air bubble relative to the vessel wall (m/s). Substitution in Equation (63) gives

$$A = \frac{24 F_g}{\pi d_b v_r T_v^2} \tag{65}$$

From Equation (62) the minimal specific surface area A_{min} is obtained

$$\text{OUR } C_x = k_{ol} A_{min} (C_o^* - C_{o,min}) \tag{66}$$

with $C_{o,min}$ being the minimum liquid oxygen concentration at which cells are able to grow.

Growth of cells in a continuous culture can be described by first order kinetics

$$C_x(t) = C_x(O) e^{k_g t} \tag{67}$$

with k_g the first order growth rate constant (s^{-1}). In order for growth of cells to occur in a sparged reactor, k_d should be sufficiently smaller than k_g:

$$k_g > k_d \tag{68}$$

For designing a continuous culture of fragile cells in a bubble column or airlift reactor, Equations (61–68) can be used. Figures 15 and 16 show the worked example for insect cells.

Inspection of the equations for k_d and A reveals that the height H of the reactor and C_o^*, thus the oxygen tension in the gas, are the important parameters to adjust to meet the demands set by the minimum specific surface area for sufficient oxygen, and by the fact that the growth rate should be faster than the death rate. The effect of the height

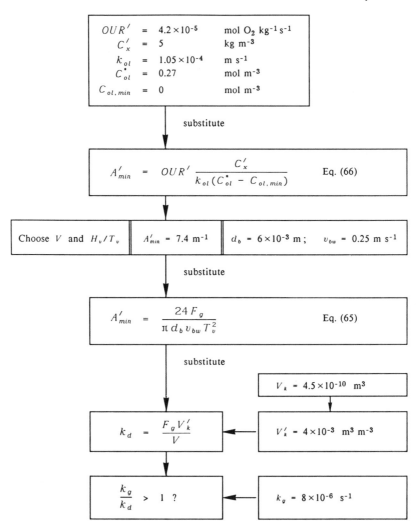

Figure 15 Bubble column design calculation scheme for growth of insect cells (adapted from Tramper et al., 1988).

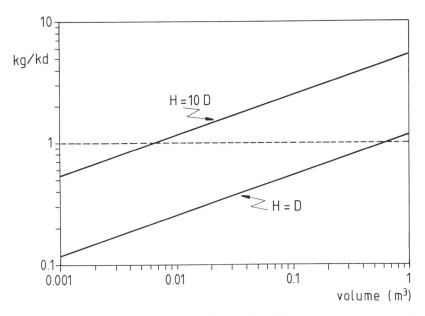

Figure 16 Double logarithmic plot of the ratio of the first order growth and death rate constant as a function of the bubble column working volume (from Tramper et al., 1988).

is clearly shown in Figure 16. In contrast, the rising velocity v_r and the air bubble diameter d_b are not easily adjustable parameters.

CONCLUDING REMARKS

It is clear that at present there is no unifying concept by which fragile cell reactors can be designed and scaled up. In this chapter we have attempted to review and evaluate various parameters that in theory could be useful for design and scale-up. We used our insect cell work as an illustration throughout even though in some cases the geometry of stirrer and vessel was not appropriate. Insect cells and animal cells can be grown well in large vessels. However, none of the theories and parameters discussed in this chapter have been validated on a larger scale than laboratory reactors. Selection of the most suitable design and scale-up method therefore needs studies in larger vessels in particular. The Kolmogorov theory and the hypothetical killing volume model are, in this respect, the promising approaches for optimal design of large-scale animal cell bioreactors.

ACKNOWLEDGMENT

We would like to thank Ir. Wim Beverloo for many fruitful discussions and for carefully reading the manuscript.

REFERENCES

Abu-Reesh, I. and Kargi, F. (1989). *J. Biotechnol., 9:* 167.

Allen, D. G. and Robinson, C. W. (1991). *Biotechnol. Bioeng., 38:* 212.

Atkinson, B. and Mavituna, F. (1983). *Biochemical Engineering and Biotechnology Handbook,* The Nature Press, New York, p. 676.

Bavarian, F., Fan, L. S., and Chalmers, J. J. (1991). *Biotechnol. Progress, 7:* 140.

Beek, W. J. and Mutzall, K. M. K. (1975). *Transport Phenomena,* Wiley, London.

Birch, J. R. and Arathoon, R. (1990). *Large-Scale Mammalian Cell Culture Technology* (A. S. Lubiniecki, ed.), Marcel Dekker, New York, p. 251.

Chalmers, J. J. and Bavarian, F. (1991). *Biotechnol. Progress, 7:* 151.

Cherry, R. S. and Papoutsakis, E. T. (1986). *Bioprocess Eng., 1:* 29.

Cherry, R. S. and Papoutsakis, E. T. (1988). *Biotechnol. Bioeng., 32:* 1001.

Cherry, R. S. and Kwon, K.-Y. (1990). *Biotechnol. Bioeng., 36:* 563.

Chisti, Y. and Moo-Young, M. (1989). *Biotechnol. Bioeng., 34:* 1391.

Croughan, M. S., Hamel, J. F., and Wang, D. I. C. (1987). *Biotechnol Bioeng., 29:* 130.

Croughan, M. S., Sayre, E. S., and Wang, D. I. C. (1989). *Biotechnol. Bioeng., 33:* 862.

Einav, S. and Lee, S. L. (1973). *Int. J. Multiphase Flow, 1:* 73.

Emery, A. N., Lavery, M., Williams, B., and Handa, A. (1987). *Plant and Animal Cells* (C. Webb and F. Mavituna, eds.), Ellis Horwood, Chichester, p. 137.

Goldblum, S., Bae, Y.-K., Hink, W. F., and Chalmers, J. (1990). *Biotechnol Prog., 6:* 383.

Handa-Corrigan, A., Emery, A. N., and Spier, R. E. (1989). *Enzyme Microb. Technol., 11:* 230.

Havenbergh, J. and Joos, P. (1983). *J. Colloid Interface Sci., 95:* 172.

Heijnen, J. J. and van 't Riet, K. (1984). *Chem. Engin. J., 28:* B21.

Hinze, J. O. (1959). *Turbulence,* McGraw-Hill, New York.

Hülscher, M., Pauli, J., and Onken, U. (1990). *Food Biotechnol., 4:* 157.

Jöbses, I., Martens, D. E., and Tramper, J. (1991). *Biotechnol. Bioeng., 37:* 484.

Katinger, H. W. D. and Scheirer, W. (1982). *Acta Biotechnologica, 2:* 3.

Kunas, K. T. and Papoutsakis, E. T. (1990). *Biotechnol. Bioeng., 36:* 476.

Levich, V. (1962). *Physico-Chemical Hydrodynamics,* Prentice-Hall, New Jersey.

MacIntyre, F. (1972). *J. Geophysical Sci., 77* (27): 5211.

Maiorella, B., Inlow, D., and Harano, D. (1988). U.S. Patent No. PCT/US88/02444.

Margaritis, A. and Zajic, J. (1978). *Biotechnol. Bioeng., 10:* 939.

Märkl, H. and Bronnenmeier, R. (1985). *Biotechnology* (H.-J. Rehm and G. Reed, eds.), Vol. 2, *Fundamentals of Biochemical Engineering* (H. Bauer, volume editor), VCH, Weinheim, p. 369.

Martens, D. E., Coco Martin, J., van der Velden- de Groot, C. A. M., Beuvery, E. C., de Gooijer C. D., and Tramper, J. (1990). *From Clone to Clinic* (D. J. A. Crommelin and H. Schellekens, eds.), Kluwer Academic Publishers, the Netherlands, p. 209.

Martens, D. E., de Gooijer, C. D., Beuvery, E. C., and Tramper, J. (1992) *Biotechnol. Bioeng. 39:* 891.

Meijer, J. J. (1989). *Effects of hydrodynamic and chemical/osmotic stress on plant cells in a stirred bioreactor,* Ph.D. thesis, Technical University Delft.

Metzner, A. B., Feehs, R. H., Ramos, H. L., Otto, R. E., and Tuthill, J. D. (1961). *A.I.Ch.E.J., 7:* 3.

Nishikawa, M. (1991). *Biotechnol. Bioeng., 37:* 691.

Oh, S. K. W., Nienow, A. W., Al-Rubeai, M., and Emery, A. N. (1989). *J. Biotechnol., 12:* 45.

Papoutsakis, E. T. (1991) *Trends in Biotechnology, 9:* 316.

Popovic, M. and Robinson, C. W. (1987). *Chem. Eng. Sci., 42:* 2825.

Popovic, M. and Robinson, C. W. (1988). *Biotechnol. Bioeng., 32:* 301.

Schlichting, H. (1979). *Boundary Layer Theory,* 7th ed., McGraw-Hill, New York.

Tramper, J., Williams, J. B., Joustra, D., and Vlak, J. M. (1986). *Enzyme Microb. Technol., 8:* 33.

Tramper, J., Joustra, D., and Vlak, J. M. (1987). *Plant and Animal Cells* (C. Webb and F. Mavituna, eds.), Ellis Horwood, Chichester, p. 125.

Tramper, J., Smit, D., Straatman, J., and Vlak, J. M. (1988). *Bioprocess Engineering, 3:* 37.

Tramper, J. and Vlak, J. M. (1988). *Upstream Processes: Equipment and Techniques* (A. Mizrahi, ed.), Alan R. Liss, New York, p.199.

van der Pol, L., Zijlstra, G., Thalen, M., and Tramper, J. (1990). *Bioprocess Engineering, 5:* 241.

van der Pol, L., Bakker, W. A. M., and Tramper, J. (1992) *Biotechnol. Bioeng., 40:* 179.

Van 't Riet, K. and Smith, J. M. (1975). *Chem. Eng. Sci., 30:* 1093.

Van 't Riet, K. and Tramper, J. (1991). *Basic Bioreactor Design,* Marcel Dekker, New York.

Wudtke, M. and Schügerl, K. (1987). *Rheologie und Mechanische Beanspruchung Biologischer Systeme,* GVC.VDI, Düsseldorf, p. 159.

Yianneskis, M., Popiolek, Z., and Whitelaw, J. H. (1987). *J. Fluid Mech., 175:* 537.

7

Serum-Free Media

Stefan A. Weiss*, **Glenn P. Godwin, Stephen F. Gorfien, and William G. Whitford**   *GIBCO BRL Cell Culture R&D, Grand Island, New York*

7.1 INTRODUCTION

Prior to 1983, insect cell culture was mainly considered to be a method for the production of viral pesticides. This consideration provided the major impetus for insect cell culture and baculovirology. A wide variety of media formulations were developed for culturing insect cells, but all needed supplementation with animal sera or other related proteinaceous materials (Weiss and Vaughn, 1986; Mitsuhashi, 1990).

In recent years, genetic engineering has revolutionized insect cell culture and baculovirus technology (Smith et al., 1983; Luckow, 1991; Agathos et al., 1990). Use of the baculovirus expression vector system (BEVS) to produce foreign gene products of medical and agricultural importance has made insect cell culture a valuable tool for biotechnology. This system has the ability to efficiently produce high levels of recombinant proteins possessing structures and biological activities closely resembling those of native proteins (Luckow, 1991). However, to develop cost-effective and consistent methods for the production of medically and agriculturally important recombinant products using insect cell culture, it is of paramount importance to eliminate animal sera from the media. Use of serum-free medium (SFM) increases cell and product yield, eliminates detrimental exogenous agents, and facilitates

Current affiliation: Calypte Biomedical, Berkeley, California downstream processing, thereby reducing the overall cost of the final product (Weiss et al., 1989; Weiss et al., 1990). Use of SFM also has advantages from a regulatory standpoint.

Serum-free media, such as SF-900 SFM, SFM-900 II SFM, and ExCell 400, have been developed for the growth of Sf-9, Tn-368 and other lepidopteran insect cells for the in vitro production of baculovirus-expressed recombinant proteins and occluded virus (AcNPV). Insect cells have been successfully adapted to and grown in serum-free medium and perform equally well in monolayer or suspension culture under both proliferative and recombinant expression modes. Large-scale production of recombinant product was accomplished in a variety of bioreactor configurations and volumes. The maximum cell yield achieved was 1.5×10^7 cells/ml in a newly designed Biospin perfusion filter bioreactor (Weiss et al., 1990). However, even though SFM was employed, there was still a need for further nutrient supplementation to allow optimal production of expressed recombinant proteins and occluded virus by cells cultured at high density (Weiss et al., 1991; Godwin et al., 1991; Jayme, 1991). Analytical studies of spent media revealed rate-limiting depletion of certain nutrient components, necessitating supplementation for maximal production of either recombinant protein or occluded virus. These rate limiting nutrients were mainly specific amino acids, carbohydrates, vitamins, lipids and sterols. When a concentrate formulated from a combination of these nutrients is added to high density cultures of insect cells grown in SFM, the yield of both recombinant proteins produced by BEVS and occluded wild type virus can be enhanced significantly. In this chapter we discuss experimental studies and developmental methods that may provide practical approaches for bench and industrial scale BEVS users. The primary emphasis will be on the development of SF-900 SFM since the authors are most familiar with this medium.

7.2 ROLE OF INSECT CELL CULTURE MEDIA IN BIOTECHNOLOGY

A crucial factor in cell culture technology is the formulation of media (Weiss et al., 1990, 1991; Godwin et al., 1991; Jayme, 1991; Weiss et al., 1980). Cell culture media demand is presently very selective. Research, development and manufacturing by pharmaceutical, veterinary and biotechnology industries demand SFM and preferably media that are chemically defined. There are well-justified reasons for such demands; animal sera is a problematic supplement for culturing cells and for manufacturing products. It is the most variable, uncharacterized, and uncontrolled component of culture medium, negatively affecting downstream processing, recovery, and therefore the final cost of the product (Weiss et al., 1980, 1990).

All cultured insect cell lines up to now have been initiated and established in media fortified with fetal bovine sera and other uncharacterized additives (Vaughn and Fan, 1989). Examples of advantages and disadvantages of insect versus mammalian cells used in process development for biotechnological applications are provided in Table 1. Most of

Table 1 Comparison of Insect and Mammalian Cells in Culture: Selection of Properties for Process Development Technology

Properties	Insect	Mammalian
Growth in serum-free media	Easy	Difficult
Growth to high density	Easy	Difficult
Dependence on proteinaceous components in the medium	No	Yes
CO_2 dependence	No	Yes
Cell maintenance	Very easy	More difficult
Immortality	Yes	Selective (transformed lines only)
Suspension/adherence versatility	Yes	No
Contact inhibition	Minimum or absent	Present (lymphoid and transformed cells are exceptions)
Dissociating agents	Not required	Required
Recombinant protein expression	High	Low
Incubation temperature	28°C	37°C
Susceptibility to changes in pH, DO, osmotic pressure	Low	High
FDA approval of biotherapeutics from cellular substrates	Yes	Yes
Potential for pathogenicity of viral vectors to humans	No	Yes

the inherent properties of insect cells applicable to process development for therapeutic products are very desirable. The fact that SFM can be used for the growth and maintenance of insect cells makes them even more attractive. Therefore, selection of an appropriate SFM is of paramount importance when establishing processes and applications implementing BEVS or procedures for occluded virus production.

7.3 SELECTION OF MEDIA FOR BACULOVIRUS EXPRESSION VECTOR SYSTEM

There are currently about 400 insect cell lines which have been established in the last 25 years, and about 60 types of insect cell culture media have been reported (Mitsuhashi, 1982; Hink et al., 1985; Agathos et al., 1990; Hink and Bezanson, 1985). In recent years, more attention has been given to the development of insect cell culture media for the production of recombinant proteins using BEVS. Interest in this field is documented by recent scientific reports in some 400 publications (Luckow, 1991). There are approximately 2,000 laboratories in various biotechnology and pharmaceutical companies that are in the process of using insect cells for feasibility studies, process development, or product manufacturing (Agathos et al., 1990; Vaughn and Weiss, 1991b).

Table 2 lists types of existing SFM that are relevant to BEVS technology. After evaluating these media for process development and pilot-scale application, most of these formulations are found to present a variety of potential disadvantages. They are either ill defined, too rich in protein, contain expensive serum substitutes, do not support cell growth to economically desirable levels, or are problematic for downstream processing and final product recovery. Successful virus replication and recombinant protein expression depends on optimal cellular physiologic processes. For this reason, insect cell culture medium is an extremely important component of the system. Ideally, the in vitro environment for cellular physiology should resemble as closely as possible that which exists in vivo. Therefore, the ideal approach would be to formulate the insect media based on the composition of insect hemolymph (Vaughn and Weiss, 1991a).

Previously, successful cell growth and recombinant protein production were demonstrated in protein-free SFM (Weiss et al., 1989). In those studies, it was demonstrated that there was no need for animal-derived proteins as media supplements. In fact, when properly delivered to the cells, the appropriate SFM formulations containing biological lipids and yeastolate devoid of proteins create nearly ideal conditions for production of recombinant proteins or occluded virus in baculovirus

Table 2 Serum-Free Media for Insect Cells (Relevant to BEVS Technology)

Media Name	Reference	Serum replacement	Applications
SFM-5	Vaughn and Weiss, 1991	LAH, yeastolate bactotryptose, bactopeptone	Tn-368
CDM	Weiss et al., 1981	Trace metals, fatty acids, sterols	*A. aegypti, A. gambil, S. frugiperda,* and AcNPV replication
Roder SFM	Roder, 1982	Egg yolk emulsion, glucose tryptose and Grace's (1962) vitamins and amino acids	*S. frugiperda, M. brassicae,* and *S. litoralis*
IZD-LPD2	Hink & Bezanson, 1985	LAH, yeast extract, methylcellulose, trace metals, cholesterol	*M. brassicae, S. frugiperda,* and AcNPV replication
Protein-free insect cell culture media	Wilkie & Stockdale, 1980 Fraser, 1989	IPL-41 basal media F-68 lipid emulsion yeastolate UF 10,000 MW	Sf-9 and rMCSF production

infections. These studies were confirmed using SFM to study the scale-up of insect cells to high density (exceeding 1.4×10^7 cells/ml) in a spin filter perfusion bioreactor (Weiss et al., 1990).

Currently, there are two commercially available serum-free and protein-free insect cell culture media used in recombinant DNA technology. These have applications for research, process development and expression of recombinant genes by BEVS. Table 3 lists the commercially available SFM for insect cell culture as well as applications for each formulation.

For scale-up purposes, the most important aspect of SFM formulation is the ability to maintain consistency in the production and performance of large lots. These parameters must be carefully monitored by an expert quality control department using an appropriate indicator cell line for bioassay. It is desirable, if not mandatory, to perform the qualifying bioassay using the cell line most widely used for foreign gene expression by BEVS. Another important aspect of the SFM formulation is the

Table 3 Commercially Available Protein-Free Media for Insect Cells Used by rDNA/Technology for Research and Production

Media Name	Application
SF-900 II (GIBCO/LTI)	1. Sf-9, Sf-21, Tn-368, *Lymantria dispar,* and *Drosophila melanogaster* cells 2. Protein expression for various fused and non-fused rDNA products, wildtype AcNPV 3. Bioreactors: Celligen, and Braun stirred tanks, Chemap airlift, and BioSpin Perfusion
Protein-free insect cell culture media (GIBCO/LTI) IPL-41, yeastolate 10,000 MW, UF (50X), LPDS (100X), F-68 (100X)	Same as SF-900 II
EX-CELL 400 (JRH)	1. Sf-9 cells 2. Protein expression for fused and nonfused rDNA products 3. Bioreactors: airlift and stirred tanks

stability of the serum substitutes such as lipids and antioxidants. These must be adequately and uniformly dispersed and/or emulsified so as to remain stable, without precipitation, for at least six months' storage at 4°C. Addition of surfactants is extremely important to SFM formulations because the cells grown in SFM tend to be more sensitive to shear stress, especially in the conditions of suspension culture bioreactors for large-scale manufacturing. There are a variety of surfactants and shear protective agents which have been successfully employed in cell culture (Murhammer & Goochee, 1990; Weiss et al., 1982). However, the user of SFM needs to be aware that some surfactants may interfere with downstream processing of the desired recombinant product.

Figure 1 shows a scheme for the formulation of serum-free media for insect cells. The main components contributing to acceptable cell yield and virus replication in SFM have been previously reported (Weiss et al., 1989; Mitsuhashi, 1982; Agathos et al., 1990; Roder, 1982; zu Putlitz et al., 1990). However, to optimize media composition for a specific process, it should be appreciated that some ingredients may be needed for optimum virus replication and/or recombinant gene expression, rather than for cell growth (Weiss et al., 1982). In general, a typical basal medium for establishing SFM contains optimized buffers, inor-

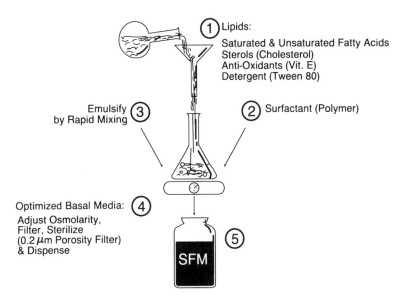

Figure 1 Formulation of Serum-Free Medium for Insect Cell Culture.

ganic and organic salts and acids, sugars, trace metals, amino acids and vitamins (Weiss et al., 1989). To maximize performance of SFM, the basal formulation needs to be fortified with lipids, increased organic acids, antioxidants, vitamins (particularly water soluble vitamins), sterols (e.g., cholesterol), membrane precursors (e.g., monothiogly-cerol and selected detergent-like molecules) and nonionic polymer molecules (e.g., Tween 80, Pluronic polyols, etc.). The users of SFM should also be aware that lipids are very sensitive to the effects of oxygen radicals despite the presence of oxygen scavengers and antioxidants, such as ethanolamine and vitamin E. It is suggested that once a bottle containing SFM has been opened and partially used, the head space should be purged with nitrogen to displace oxygen before storage. Likewise, during filtration in the manufacturing of SFM, large formulation tanks containing media should be continuously stirred and the head space purged with clean nitrogen. Emulsification or dispersion of the lipids and related components is also an important step in media formulation. A manual process may suffice for small laboratory-scale lots involving short-term research with rapid turnover of the media. However, for large-scale commercial purposes (in either research or manufacturing), emulsification or dispersion of the lipid complex should be performed using an advanced medical grade stainless steel automated system pressurized with nitrogen.

7.4 ESTABLISHING CELL GROWTH KINETICS IN SFM

The most practical and versatile method of growing insect cells in suspension is in a 100 ml shake flask. This method is not oxygen-limited and provides ideal in vitro conditions for cell growth nutritionally optimized media.

Figure 2 illustrates the growth kinetics of *Spodoptera frugiperda* (Sf-9) cells in two commercially available SFM formulations, GIBCO SF-900 SFM and ExCell 400 SFM from JRH Biosciences. Cells were planted at 2.0×10^5 cells/ml in 250ml plastic Erlenmeyer flasks (Corning #25600-250) with loosened caps to facilitate oxygenation. The cultures were placed on an orbital shaker platform at 125–135 rpm at $28 \pm 0.5°C$ in ambient atmosphere without additional CO_2. Cell counts were carried out daily until 10 days post planting. The cell growth characteristics in both media were similar.

In our previous studies it was demonstrated that concentrations of specific amino acids decreased in the spent medium as the cultures approached high densities (Weiss et al., 1990; Vaughn and Weiss, 1991a).

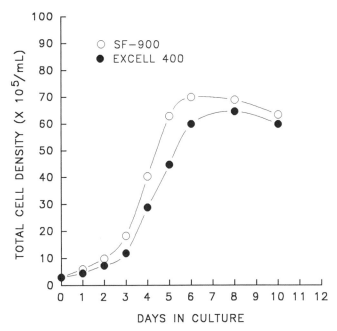

Figure 2 Comparison of Sf-9 Cell Growth in SF-900 SFM vs. ExCell 400. Sf-9 cells were planted at a density of 3×10^5 cells/ml in 250 ml shake flasks (100ml volume) on an orbital shaker platform rotating at 125–135 rpm in either GIBCO SF-900 SFM or JRH ExCell 400 SFM. The cells were incubated at 27.5°C and samples taken daily for quantitation of total cell density. Cells grown in SF-900 SFM reached a higher peak cell density than cells grown in ExCell 400.

Amino acids that were depleted in high density culture were cysteine, leucine, serine, threonine, tyrosine, valine and glutamine. The most significant decrease was observed in the concentration of glutamine, which on day 4 post planting was only about 50% of the original level, and on day 6 in culture was completely exhausted. We then began to investigate the effect of replacement of the depleted amino acids as well as sugars on cell growth in SFM.

Figure 3 depicts the growth of Tn-368 cells in serum-free and serum-supplemented media. In our hands, Tn-368 cells were more difficult to grow in SFM than Sf-9 cells. We therefore chose to investigate whether or not replacement of the identified depleted nutrients in SFM would improve the proliferation of Tn-368 cells in shaker culture. SF-900 SFM, when supplemented with additional quantities of the nutrients shown to

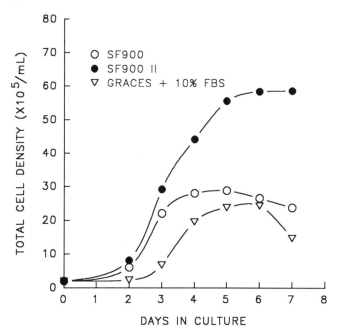

Figure 3 Growth kinetics of Tn-368 in SF-900 SFM vs. SF-900 II SFM or Grace's plus 10% FBS. Tn-368 cells were planted at a density of 2.0 × 10⁵ cells/ ml in 250ml shake flasks on an orbital shake platform rotating 125–135 rpm in SF-900 SFM, SF-900 II SFM, or Grace's (supplemented with 10% FBS growth media. The cells were incubated at 27.5°C and samples taken daily for quantitation of total cell density. Cells grown in SF-900 II SFM reached a peak cell density almost twofold higher than cells grown in either SF-900 or Grace's supplemented 10% FBS.

be depleted first, resulted in the new formulation, SF-900 II SFM. As shown in Figure 3, a dramatic increase in Tn-368 cell growth can be seen when comparing growth in SF-900 II SFM with SF-900 SFM and Grace's TNM-FH plus 10% FBS. Cells grown in SF-900 SFM reached a maximum density of 2.8×10^6 viable cells/ml on day 4 post planting and maintained this plateau for an additional 72 hr. Cells grown in Grace's TNM-FH plus 10% FBS reached maximum cell density of 2.25×10^6 viable cells/ml on day 5 post planting. However, the cells in SF-900 II SFM reached maximum density of about 6.0×10^6 viable cells/ml on day 5 post planting and maintained the same plateau for an additional 48 hours. This provides an added value for processes which require longer periods of viability, such as virus production and expression of recombinant genes by BEVS.

7.5 OPTIMIZATION OF RECOMBINANT GENE EXPRESSION IN INSECT CELL CULTURE USING SFM

Our next objective was to investigate whether or not the production of recombinant proteins could be increased in batch suspension shaker cultures grown in SFM. Figure 4 summarizes the results of these experiments. Sf-9 cells were planted at 2×10^5/ml in shaker cultures using

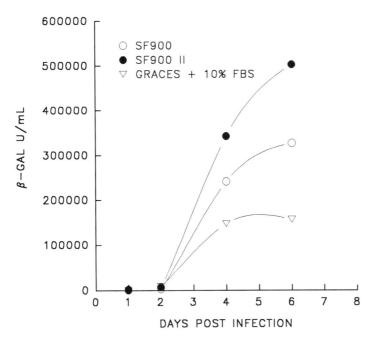

Figure 4 Expression of recombinant β-galactosidase (rβ-GAL) in Sf-9 cells grown in various media. Sf-9 cells were planted at a density of 3.0×10^5 cells/ml in 250ml shake flasks (100ml volume) on an orbital shaker platform rotating at 125–135 rpm in SF-900 SFM, SF-900 II SFM or Grace's with 10% FBS growth media. On day 3 post planting the cultures were infected with rAcNPV (clone VL-941) at a MOI of 0.5. The cultures were sampled post infection for assay of rβ-GAL activity. Sf-9 cells in SF-900 II SFM supported better rβ-GAL expression than cells grown in either SF-900 SFM or Grace's (Supplemented) + 10% FBS.

various insect cell culture media. When the cell densities reached about 2 × 10^6/ml, the cultures were centrifuged, resuspended in fresh media, and infected at a multiplicity of infection (MOI) of 0.5 with a rAcNPV (VL-941) engineered to express recombinant β-galactosidase. The supernatants were assayed for β-galactosidase activity on days 4 and 6 post infection using a previously established assay method (Weiss et al., 1990). A maximum of 500,000 units/ml of recombinant β-galactosidase was produced by shaker cultures grown in SF-900 II SFM 6 days post infection.

Overall, the production of recombinant β-galactosidase by Sf-9 cells grown in SF-900 II SFM in shaker culture was 153% of the product obtained from the culture grown in SF-900 SFM. A significant result of this experiment was the dramatic increase in yield (330%) of recombinant β-galactosidase produced in SF-900 II over the yield of this product in Grace's TNM-FH supplemented with FBS. It should be noted, however, that these cultures were fed at the time of infection, which provides optimal conditions for expression in the original SFM and serum-supplemented cultures.

Previously, we demonstrated production kinetics for recombinant erythropoietin (rEPO) by BEVS using SFM (Weiss et al., 1990). In our most recent studies, we investigated production of rEPO in Sf-9 cells grown in a 5 liter Celligen bioreactor (New Brunswick Scientific) operated in batch mode using SFM (Figure 5). Sf-9 cells from an exponentially expanding culture were planted in 3.5 liter of SF-900 II at 2 × 10^5 viable cells/ml. When the viable cells in the bioreactor reached a density of 2.7 × 10^6/ml on day 4, the culture was infected with rAcNPV at a MOI of 0.5. The bioreactor was sampled daily post infection to determine rEPO levels. The level of glucose was also monitored daily post infection. At the final harvest on day 3 post infection, there was 70% of the original level of glucose remaining, indicating this was not a rate-limiting factor. The yield of rEPO from Sf-9 cells grown in this batch culture bioreactor using SF-900 II SFM reached a maximum level of 3,300 units/ml. A total of 84 mg of rEPO was recovered from this experiment.

To determine if specific components, such as amino acids and other identified nutrients (Weiss et al., 1990) at particular concentrations are major contributors to improved Sf-9 and Tn-368 cell growth and recombinant protein expression in SF-900 II SFM, we formulated these components into a separate concentrated supplement or "enhancer." We then tested this enhancer by adding it to a culture of Sf-9 cells in SF-900 SFM and infecting with a rAcNPV expressing β-galactosidase. Sf-9 cells were planted at a density of 2 × 10^5 viable cells/ml and grown to 3 × 10^6 viable cells/ml in 5 days. Enhancement supplement was then added to these

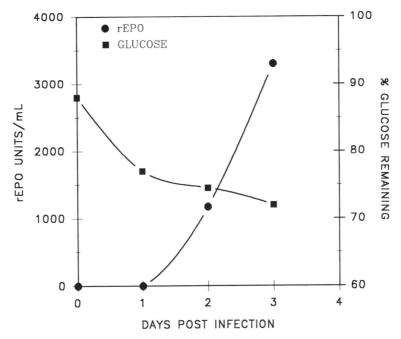

Figure 5 Kinetics of rEPO production in bioreactor using SF-900 II serum-free medium. Sf-9 cells were grown in a 5 liter Celligen bioreactor using GIBCO SF-900 II SFM. When the cell density reached 2.7×10^6 cells/ml (day 4 post planting), the culture was infected with a rAcNPV at an MOI of 0.50. A total of 1.3×10^7 units or 84 mg of rEDO was produced. Glucose was utilized 30% by day 3 post infection and was not a rate limiting factor.

shaker cultures. One hour later, cultures were infected with rAcNPV (VL-941) at a MOI of 0.5. At 5 days post infection, the cultures were harvested and each supernatant was assayed for β-galactosidase activity. Table 4 depicts the stimulation of β-galactosidase production using SFM plus enhancer. Cultures grown in SF-900 SFM, then supplemented with enhancer and infected with rAcNPV, produced 300% more recombinant β-galactosidase than cultures grown and infected in SF-900 SFM without enhancer.

CONCLUDING REMARKS

The use of insect cell culture and related technologies is no longer limited to academic research as an in vitro model for the examination of novel

Table 4   Enhancement of β-Galactosidase Production by BEVS Using SFM[a]

Media variations	Cell density (P.I.)[b]		β-galactosidase activity units/ml	% Increase
	Day 2	Day 5		
SF-900 control	6.0	0	100, 500	
SF-900 + enhancer	5.0	2.00	400, 300	300

[a]Sf-9 cells planted at 2×10^5 cells/ml were grown to 3.0×10^6 cells/ml (5 days) and were supplemented with the enhancer (see text) and incubated for one hr. Following incubation, the cultures were infected with rAcNPV (VL-941, MOI = 0.5). Five days post infection the cultures were harvested, cells separated, and supernatant assayed for β-galactosidase activity.
[b]Viable cell density ($\times 10^6$) post infection (P.I.)

recombinant constructs and expression techniques. Currently, insect cells cultured in a variety of systems have many uses in agriculture, forestry, and medicine. The rapidly increasing trend to employ BEVS for the expression of foreign genes in insect cells has established a need for serum-free media development for use in large-scale expression systems.

The scientific literature is replete with information regarding the requirement of specific components for cell proliferation and virus replication. Now, however, there is an interesting need to understand the function of specific media components in regulating cellular proliferation and support, viral replication, and foreign gene expression in cells infected with various baculovirus strains and constructs.

We have examined two model nonfusion vectors expressing recombinant proteins (β-galactosidase and EPO) having greatly different molecular weights and other characteristics. By using analytical techniques such as HPLC and FPLC, we were able to quantitate the depletion of specific amino acids, carbohydrates and lipids in high density insect cell culture. When these depleted nutrients were added to SFM, a significant improvement was observed. This improved SFM allowed increased proliferation of both Sf-9 and Tn-368 cells, and enhanced production of recombinant β-galactosidase and rEPO by BEVS.

When rEPO produced in a 5 liter Celligen bench-scale bioreactor using SFM was purified to homogeneity and analyzed by western blot and a function-specific bioassay, it was found to be sufficiently glycosylated and closely resembled the native protein in every respect. We have demonstrated that the system can be scaled up, suggesting that this process may be applicable for manufacturing recombinant proteins for use in a variety of biotechnological and biomedical products.

For optimum expression of recombinant proteins produced in high density insect cell cultures, those nutrients shown to be depleted during growth and expression either need to be replaced in the SFM media during infection, or added into the initial formulation prior to planting the cells. Considerable progress has been made to date in optimizing SFM formulations for insect cell culture. However, additional biochemical definition of specific nutrients depleted in high density cultures should expedite the formulation of a more cost-effective SFM for many purposes, including the large-scale manufacture of biologicals. Identifying the specific nutrients depleted in high density serum-free insect cell culture may be a critical step toward the realization of an efficient and productive biotechnological process.

REFERENCES

Agathos, S. N., Jeong, Y. H., and Venkat, K. (1990). Growth kinetics of free and immobilized insect cell cultures, *Ann. N.Y. Acad. Sci., 589:*372–398.

Fraser, M. J. (1989). Expression of eucaryotic genes in insect cell cultures, *In Vitro Cell. and Dev. Biol., 25:*225–235.

Godwin, G., Gorfien, S., Tilkins, M. L., and Weiss, S. A. (1991). "Development of a Low Cost Serum-Free Medium for the Large-Scale Production of Viral Pesticides in Insect Cell Culture," In: Proc. Eighth International Conference on Invertebrate Fish and Tissue Culture, pp. 102–110, (M. J. Fraser, Jr. ed). Tissue Culture Association, Columbia, MD.

Hink, W. F., Ralph, D. A., and Joplin, K. H. (1985). *Comprehensive Insect Physiology Biochemistry and Pharmacology* (G. A. Kerkut and L. I. Gilbert, eds.), Pergamon Press, New York, p. 547.

Hink, W. F. and Bezanson, D. R. (1985). *Techniques in Life Sciences* (E. Kurstak, ed.), Elsevier Scientific Publishers, Ireland Ltd. C111:19

Inlow, D., Harano, D., and Maiorella, B. (1987). 194th ACS Meeting, Division of Microbial and Biochemical Technology, abstract 70.

Jayme, D. W. (1991). *Cytotechnology, 5:*15–30.

Luckow, V. A. (1991). *Cloning and Expression of Heterologous Genes in Insect Cells with Baculovirus Vectors, Recombinant DNA Technology and Applications* (A. Prokop, R. Bajpai, and C. S. Ho, eds.), McGraw Hill, New York pp. 97–133.

Miltenburger, H. G. (1983). European Conference on Hormonally Defined Media, 1st Issue, pp. 31–43.

Mitsuhashi, J. (1982). *Advances in Cell Culture* (K. Maramorosch, ed.), Academic Press, New York, p. 133.

Mitsuhashi, J. (1990). *Invertebrate Cell System Applications,* (J. Mitsuhashi ed.), CRC Press, Boca Raton, Florida, p. 3.

Murhammer, D. W. and Goochee, C. F. (1990). *Biotechnol. Prog.,* 6:142–148.

Roder, A. (1982). *Naturwissensch 69:*92–93.

Smith, G. E., Summers, M. D., and Fraser, M. J. (1983). *Mol. Cell. Biol., 3:*2156–2165.

Vaughn, J. L., and Fan, F. (1989). *In Vitro, 25:*143–145.

Vaughn, J. L., and Weiss, S. A. (1991a). *BioPharm 4:*16–19.

Vaughn, J. L., and Weiss, S. A. (1991b). *Large Scale Mammalian Cell Culture Technology* (A. Lubiniecki, ed.), Marcel Dekker, Inc., New York, p. 597.

Weiss, S. A., Lester, T. L., Kalter, S. S., and Heberling, R. L. (1980). *In Vitro, 16:*616–628

Weiss, S. A., Smith, G. C., Kalter, S. S., and Vaughn, J. L. (1981a). Improved method for the production of insect cell cultures in large volume, *In Vitro, 17:*495–502.

Weiss, S. A., Smith, G. C., Kalter, S. S., and Vaughn, J. L. (1981b). *In Vitro 17:*495–502.

Weiss, S. A., Smith, G. C., Vaughn, J. L., Dougherty, E. M., and Tomkins, G. J. (1982). Effect of aluminum chloride and zinc sulfate on *Autographa californica* nuclear polyhidrosis virus (AcNPV) replication in cell culture, *In Vitro, 18:*937–944.

Weiss, S. A. and Vaughn, J. L. (1986). Cell culture methods for large-scale propagation of baculoviruses, *The Biology of Baculoviruses, Vol. II* (R. R. Granados and B. A. Federici, eds.), CRC Press, Boca Raton, Florida, p. 64.

Weiss, S. A., Belisle, B. W., DeGiovanni, A., Godwin, G., Kohler, J., and Summers, M. D. (1989). "Insect Cells as Substrates for Biologicals," Proceedings of the International Association for Biological Standardization, Symposium on Continuous Cell Lines as Substrates for Biologicals, Arlington, Virginia, pp. 271–289.

Weiss, S. A., Gorfien, S., Fike, R., DiSorbo, D., and Jayme, D. (1990). "Large Scale Production of Proteins Using Serum-Free Insect Cell Culture," Proceedings of Ninth Australian Biotechnology Conference in Biotechnology: The Science and the Business, Gold Coast, Queensland, Australia, pp. 220–231.

Weiss, S. A., DiSorbo, D. M., Whitford, W. G., and Godwin, G. (1991). "Improved Production of Recombinant Proteins in High Density Insect Cell Culture," In: Proc. Eighth International Conference on Invertebrate and Fish Tissue Culture, pp. 153–159, (M. J. Fraser, Jr. ed). Tissue Culture Association, Columbia, MD.

Wilkie, G. E. I. and Stockdale, H. (1980). Proceedings of the Third General Meeting of the European Society of Animal Cell Culture Technology (Griffith B. W. Hennessen, F. Horodniceame, R. Spier, eds.) S. Karger, pp. 29–37.

Zu Putlitz, J., Kubasek, W. L., Duchene, M., Margaret, M., von Specht, B. U., and Domdey, H. (1990). Antibody production in baculovirus-infected cells, *Biotechnology, 8:*651.

8

Foreign Gene Expression in Insect Cells

Donald L. Jarvis   *Texas A&M University, College Station, Texas*

8.1. INTRODUCTION

A major reason for the recent intensification of interest in insect cell culture is that insect cells have emerged as an important eucaryotic host for the expression of recombinant gene products. Lepidopteran insect cells serve as the in vitro hosts for baculoviruses, which are one of the most widely used expression vectors currently available. One can construct a recombinant baculovirus that contains the foreign gene of interest, add the recombinant virus to an insect cell culture, and obtain high level expression of the foreign gene product. Stable transformation is another approach that has been developed for the expression of a foreign gene product in insect cells. A cloned gene can be placed under the control of an appropriate transcriptional promoter, added to insect cells, and a cell subpopulation can be selected that contains the new gene and will express it continuously, in the absence of viral infection.

This chapter will cover the use of insect cell cultures for foreign gene expression. The baculovirus expression vector system and the transformed insect cell approach both will be described. The capabilities and limitations of these two different methods will be discussed and the specific advantages and disadvantages of each will be considered. Further details on baculoviruses and their use as expression vectors can be obtained by reading Granados and Federici (1986), reviews by Kang (1988), Luckow and Summers (1988a), Miller (1988), Vlak and Keus (1990), Blissard and Rohrmann (1990), and Chapter 2 of this volume.

8.2 BACULOVIRUS-MEDIATED FOREIGN GENE EXPRESSION IN INSECT CELLS

Replication of Baculoviruses

The Baculoviridae are a large family of enveloped, double stranded DNA viruses that infect only invertebrates, mainly insects (reviewed in Granados and Federici, 1986). The family contains three subgroups. Two are distinguished by their ability to produce viral occlusions, or polyhedra, in the nuclei of infected insect cells. Viral occlusions are large, paracrystalline particles that consist of progeny virions embedded within a dense, proteinaceous matrix. This matrix protects the virions from inactivation and allows baculoviruses to remain infectious for many years in the environment. A baculovirus infection begins when an insect eats food that is contaminated with occluded virus particles. The strongly alkaline conditions of the insect midgut cause the viral occlusion to dissociate, and infectious virions are released. The enveloped virions then can fuse with midgut epithelial cells and migrate to the nucleus, where they uncoat and replicate. The replication cycle of the subgroup A baculoviruses is biphasic, and distinct types of viral progeny are produced during each phase. In the first phase, which occurs relatively soon after infection, nonoccluded nucleocapsids are assembled in the nucleus and these subsequently migrate to the plasma membrane and escape as enveloped virions by budding from the cell. These extracellular virions disseminate the infection from the midgut to other tissues within the individual infected insect. In the second phase of viral replication, which occurs later during the infection, occluded virions are produced in the nucleus. These are released into the environment when the infected insect dies, where they can transmit the infection to other insects even after lying dormant for many years, as mentioned above.

At the molecular level, the baculovirus replication cycle can be divided into four different phases: immediate early, delayed early, late, and very late (Friesen and Miller, 1986; Guarino, 1989). Viral genes are expressed in a sequential, temporally regulated fashion during one or more of these phases. A cascade type of transcriptional regulation is probably responsible for this mode of gene expression. That is, the immediate early genes are expressed immediately after infection, in the absence of other viral functions, and one or more of the resulting gene products induces transcription of the delayed early genes. Some delayed early gene products then induce transcription of the late genes, and finally, the very late genes are expressed under the control of gene products expressed during the previous phases. One relatively well-defined component of the baculovirus regulatory cascade is IE1, an immediate early gene discovered in *Autographa californica* (multicapsid) nuclear polyhedrosis virus (Ac*M*NPV; Guarino and Summers, 1986a, 1987). The IE1 gene is expressed in the absence of de novo viral protein synthesis and it encodes a protein that transactivates, among others, the delayed early 39K gene. Two other well-characterized viral genes are polyhedrin, a very late gene that encodes the major component of the viral occlusion particle (Smith, Vlak, and Summers, 1983; Hooft van Iddekinge, Smith, and Summers, 1983), and p10, another very late gene that appears to function in assembly of the viral occlusion particle (Smith, Vlak, and Summers, 1983; Kuzio et al., 1984; Vlak et al., 1988).

Baculovirus Expression Vectors

The protective matrix of a viral occlusion particle consists mainly of a single protein, polyhedrin, which is produced in massive amounts during a baculovirus infection. One reason for this is that the viral polyhedrin gene contains an unusually strong transcriptional promoter, which induces the synthesis of extremely large amounts of mRNA. The ability of baculoviruses to produce extremely large amounts of polyhedrin mRNA and protein was the critical molecular basis for their development as expression vectors. It was predicted that if large amounts of polyhedrin could be produced under the control of this strong promoter, then it could be used for the high expression of other genes, as well. Another important fact was that the polyhedrin protein is not essential for the replication of baculoviruses in cultured insect cells (Smith, Fraser, and Summers, 1983). In the absence of polyhedrin, occluded virus cannot be produced, but the nonoccluded, extracellular form of the virus is produced in normal amounts.

Thus, the classical baculovirus expression vector may be defined simply as a recombinant insect virus in which the sequences encoding polyhedrin have been replaced with sequences encoding a chosen foreign gene product. When susceptible insect cells are infected with this virus, the newly inserted gene can be expressed under the control of the polyhedrin promoter. The foreign gene will be highly transcribed in place of polyhedrin and the gene product will usually accumulate in large amounts in the infected insect cell culture. Analogous expression vectors have been produced in which the foreign gene has been placed under the control of the p10 promoter, derived from another highly expressed, nonessential baculovirus gene (Vlak et al., 1988). Other vectors contain duplicated promoter sequences and each individual promoter element controls the expression of a different foreign gene product (Emery and Bishop, 1987). In all of these vectors, the strength of the promoter is the most significant advantage over other eucaryotic expression vectors. Large amounts of mRNA are synthesized and the recombinant protein will often constitute well over half of the total protein in the infected cell by the very late phase of infection.

Construction of a Baculovirus Expression Vector

The classical method for construction and isolation of a baculovirus expression vector is outlined in Figure 1 (Summers and Smith, 1987). First, a cloned copy of the desired coding sequence is inserted into a transfer vector (Figure 1, step 1). A variety of transfer vectors have been constructed, but each basically consists of a bacterial plasmid (pUC) that contains the polyhedrin promoter, long 5′ and 3′ flanking sequences, and a multiple cloning site for insertion of the desired coding sequence. Almost all of the polyhedrin coding sequence has been deleted in these plasmids. The most effective transfer vectors are specifically designed so that the foreign gene is inserted downstream of the first nucleotide in the polyhedrin initiation codon. Studies from several laboratories have shown that this provides the highest levels of foreign gene expression, by preserving the transcriptional and translational control sequences located in the 5′ noncoding region of the polyhedrin gene (Matsuura, Possee, and Bishop, 1987; Luckow and Summers, 1988b, 1989; Possee and Howard, 1987; Rankin, Ooi, and Miller, 1988). More details on many different transfer vectors can be obtained in the preceding references and in the reviews by Luckow and Summers (1988a) and Miller (1988).

After insertion of the foreign coding sequence, the recombinant gene must be incorporated into the baculovirus genome. This is necessary because the polyhedrin promoter cannot express anything by itself: like all other very late genes, it requires an unknown number of other viral functions for its transcriptional activity. The recombinant gene is incorporated into the viral genome by cotransfecting insect cells with a mixture of wild type viral DNA and transfer vector plasmid DNA (Fig-

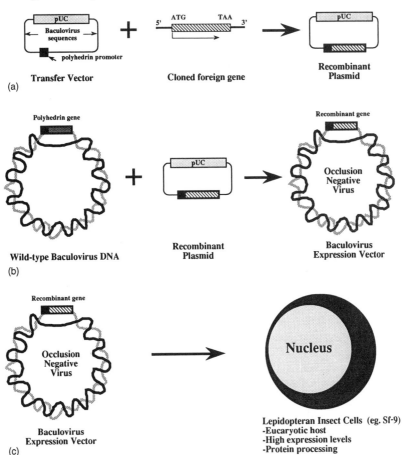

Figure 1 Preparation of a recombinant baculovirus for foreign gene expression. The procedure used for the preparation of a classical baculovirus expression vector and its use for foreign gene expression are shown. Details are given in the text.

ure 1, step 2). The flanking sequences in the plasmid target the recombinant gene to the viral polyhedrin locus, where genetic exchange occurs by homologous recombination. This generates a small percentage of recombinant viruses in which the wild type polyhedrin gene has been replaced by the foreign gene. The mixture of recombinant and wild type viral progeny is then resolved in a conventional plaque assay (Hink and Vail, 1973; Volkman and Summers, 1975). The recombinants in which the polyhedrin coding sequence has been replaced by a foreign coding sequence will produce occlusion negative plaques, while wild type viruses will produce occlusion-positive plaques. Occlusion-positive and -negative plaques can be distinguished under the light microscope and, on this basis, recombinants can be visually identified, picked, and amplified. Once a working virus stock is obtained, it can be used to infect cultured insect cells and its ability to express the foreign gene product can be assessed (Figure 1, step 3).

Capabilities of Baculovirus Expression Vectors

Expression Levels

As mentioned above, the main advantage of the baculovirus expression system is that it usually provides extremely high levels of foreign gene expression. The levels of expression obtained for various gene products have been estimated at about 1 to 500 mg of the recombinant protein per liter of infected insect cells (about 2×10^9 cells; Luckow and Summers, 1988a). The level of foreign gene expression that one actually achieves in the baculovirus system probably depends upon a large number of factors, but only a few of these are well defined. One is the type and quality of insect cells used as hosts for the viral infection. A commonly used cell line is IPLB-Sf21-AE (Sf-21), originally established from ovarian tissue of the fall armyworm, *Spodoptera frugiperda* (Vaughn et al., 1977). Another is Sf-9, a clonal derivative of Sf-21 (Summers and Smith, 1987). High quality growth medium and routine subculturing are necessary to maintain these cells in a healthy state. For optimal foreign gene expression, the cells must be in the log phase of growth and have a viability of at least 97% at the time of infection (Summers and Smith, 1987). Historically, it has been difficult to maintain high cell viability in larger insect cell cultures (i.e., > 1 liter). These cultures must be sparged to accommodate the unusually high oxygen demand of the cells, but this can induce physical damage, because the cells are highly sensitive to shear stresses induced by the sparging process (Weiss et al., 1980, 1982). Recently, this problem was solved with the discovery that the cells in a large-scale insect cell culture can be protected from shear stress by including a nonionic surfactant, Pluronic F-68, in the growth medium (Maiorella et al., 1988; Murhammer and Goochee, 1988).

The use of different insect cell lines for baculovirus-mediated foreign gene expression was recently examined in an extensive study by Hink and coworkers (1991), who expressed three different foreign genes in 23 different cell lines. Generally, it was found that some cell lines provided higher expression levels than Sf-9, but no single cell line provided the highest levels of expression for all of the three products tested. In a similar study from Granados' and Wood's labs at Cornell University, the expression of two different foreign gene products was analyzed in seven different cell lines and the results showed that a cell line derived from *Trichoplusia ni* eggs (TN5B14) provided generally higher levels of expression (T. Wickham, personal communication). Thus, it appears that cell lines other than Sf-9 will be very useful for baculovirus-mediated foreign gene expression. However, it will be important to examine the expression of a large number of foreign genes in these cells, their protein processing capabilities, and their adaptability to scale-up.

A highly empirical factor that strongly influences the level of foreign gene expression obtained in the baculovirus system is the nature of the foreign gene product itself. It seems that proteins that enter the secretory pathway, as a class, are expressed at lower levels in the baculovirus system that nuclear or cytoplasmic proteins. For example, human epidermal growth factor receptor was expressed at 1 mg/liter (Greenfield et al., 1988) and human tissue plasminogen activator, including both the extracellular (0.25 mg/liter) and the intracellular (0.40 mg/liter) forms of the protein, was expressed at only 0.65 mg/liter (D. L. Jarvis, unpublished information). It must be emphasized, however, that this is a generalization that does not apply for all recombinant glycoproteins. Reasonably high expression levels have already been reported for several glycoproteins, including the cytoplasmic domain of the human insulin receptor (Ellis et al., 1988, reported 50–200 ug/10^7 cells, which is about 10–40 mg/L) and the hepatitis B virus surface antigen (Lanford et al., 1989; 90 mg/L). Furthermore, low levels of expression in the baculovirus system are often as high as the best expression levels obtained in other eucaryotic systems.

Protein Processing

A second advantage of the baculovirus expression system is that it uses a eucaryotic host that should be capable of co- and posttranslational protein processing. During the early stages of baculovirus expression vector development, however, the protein processing capabilities of lepidopteran insect cells were questioned. These cells were unfamiliar to most investigators and they were grown under rather unusual conditions, as compared to the conditions typically used for mammalian cell culture. In fact, few direct studies had been performed on the processing of proteins in lepidopteran insect cells, and, in any event, most of the recombinant proteins that were being expressed in these cells were derived from vertebrates, not from insects. Now that a large number of proteins have been expressed in the baculovirus system, extensive information has accumulated on the types of protein processing that lepidopteran insect cells can carry out. The findings can be summarized quite simply: these cells are qualitatively capable of almost every type of protein processing that has been examined, but the precise structural features of the products can differ in certain ways from the corresponding mammalian cell products. Despite these minor structural differences, every recombinant protein product that has been produced in the baculovirus system and subsequently tested for its antigenic or biological authenticity has had the properties of the native product (reviewed in Luckow and Summers, 1988a; Miller, 1988).

One processing mechanism that has been examined closely in the baculovirus-insect cell system is the proteolytic cleavage of protein precursors. Using a large number of different foreign proteins as models, including human β interferon (Smith, Summers, and Fraser, 1983) and interleukin-2 (Smith et al., 1985), it has been found that lepidopteran insect cells will cleave pre and pro sequences accurately. The internal proteolytic cleavages required for the processing of some polyprotein precursors are known to occur (Oker-Blom and Summers, 1988; Oker-Blom, Pettersson, and Summers, 1989). However, some types of internal proteolytic cleavages occur inefficiently in cultured insect cells (Possee, 1986; Kuroda et al., 1986; Rusche et al., 1987) and appear to occur more efficiently in insect larvae (Kuroda et al., 1989). This probably reflects species- or tissue-specific differences in the abilities of different hosts to carry out these types of cleavages, a possibility that emphasizes the need to examine different hosts for differences in their protein processing capabilities.

N-glycosylation is another protein processing mechanism that has been intensively scrutinized in the baculovirus system. This modification clearly occurs in insect cells, but the N-linked oligosaccharide processing pathway appears to differ from the one found in mammalian cells. The first indication of this came from structural studies on the native glycoproteins of cultured mosquito and *Drosophila* (dipteran) cells, which revealed only high mannose or trimmed high mannose side chains (Butters and Hughes, 1981; Butters, Hughes, and Vischer, 1981; Hsieh and Robbins, 1984). These studies suggested that high mannose oligosaccharide side chains could be added to insect cell glycoproteins, and the side chains could be trimmed, but they could not be converted to the complex forms commonly found in mammalian cells. Studies on baculovirus-expressed vertebrate glycoproteins suggest that this conclusion applies to cell lines derived from lepidopteran insects, such as *Spodoptera frugiperda* (Sf), as well. Glycoproteins expressed in Sf cells are usually slightly smaller than their native counterparts and this size difference can be resolved by preventing N-glycosylation or by endoglycosidase treatment (e.g., Kuroda et al., 1986; Wojchowski et al., 1987; Greenfield et al., 1988). Further, some glycoproteins can be converted to an endoglycosidase H resistant form in these cells and this can be specifically prevented by treatment with oligosaccharide processing inhibitors (Jarvis and Summers, 1989; Sissom and Ellis, 1989; Grabowski, White, and Grace, 1989). These observations indicated that high mannose oligosaccharides are trimmed to a small core structure, but are not converted to a typical complex structure. Recently, this interpretation was elegantly confirmed by direct structural analyses on the oligosaccharides isolated from an influenza hemagglutinin protein expressed in Sf-9 cells (Kuroda et al., 1990). This study revealed that the hemagglutinin molecule produced in Sf-9 cells contains high mannose type side chains or truncated side chains that contain fucose, but lacks galactose and sialic acid. In another recent study, the presence of complex type sialic acid-containing side chains was documented for the first time in an insect cell system (Davidson, Fraser, and Castellino., 1990). This was accomplished by direct structural analysis of the oligosaccharides derived from human plasminogen expressed in Sf-21 cells. In view of this finding, it will be necessary to determine the carbohydrate side chain structures of additional glycoproteins, isolated from both Sf-21 cells and its Sf-9 subclone, before we will have a clear understanding of the capabilities of the N-glycosylation pathways in each cell type.

There are other glycosylation pathways in mammalian cells, which result in the addition of oligosaccharides O-linked to serine or threonine through N-acetylgalactosamine or N-acetylglucosamine (reviewed by Hanover and Lennarz, 1981; Hart et al., 1989). The extracellular domain of the native human insulin receptor is O-glycosylated through N-acetylgalactosamine, but it was not detectably O-glycosylated when expressed in Sf-9 cells (Sissom and Ellis, 1989). The gp50 glycoprotein of pseudorabies virus, however, was O-glycosylated in Sf-9 cells (Thomsen, Post, and Elhammer, 1990). In this study, it also was reported that O-glycosylation occurred at a much lower efficiency in Sf-9 cells, probably because the cells contain relatively low levels of β-1,3-galactosyltransferase activity. It also was found that there were differences in the structures of the oligosaccharide side chains attached to gp50 in Sf-9 cells and in mammalian cells. The predominant side chain was the monosaccharide N-acetylgalactosamine, with smaller amounts of the disaccharide, galactose (β-1,3)-N-acetylgalactosamine. Terminal sialic acid residues were not detected. Finally, recent studies on a 42 kD structural polypeptide of the AcMNPV viral occlusion particle have provided the first evidence that Sf cells can add O-linked N-acetylglucosamine (Whitford and Faulkner, 1990). This modification occurs by a pathway that is completely distinct from the pathway that adds carbohydrates through an O-linkage with N-acetylgalactosamine (Hart et al., 1989).

Two other covalent chemical modifications that lepidopteran insect cells can carry out are phosphorylation and acylation. Several nuclear DNA binding proteins are known to be phosphorylated in Sf cells, including the human T cell leukemia virus type I $p40^X$ protein (Jeang et al., 1987; Nyunoya et al., 1988) and the simian virus 40 large tumor antigen (SVT ag; Lanford, 1988; O'Reilly and Miller, 1988). Jeang and coworkers reported that there were no differences in the phosphopeptide patterns of $p40^X$ expressed in human lymphoid or Sf-9 cells (Jeang et al., 1987). In contrast, preliminary studies in our lab indicated that SVT ag expressed in Sf-9 cells is phosphorylated at different sites than in monkey fibroblasts (D. L. Jarvis, unpublished). A more extensive study on the ability of Sf-9 cells to phosphorylate SVT ag was recently published (Hoss et al., 1990). The results showed that the phosphorylation sites recognized by nuclear kinases in mammalian cells were underphosphorylated in Sf-9 cells, while those recognized by cytoplasmic kinases were generally better utilized in Sf-9 cells.

Among the foreign proteins that are known to undergo acylation in lepidopteran insect cells are SVT ag (Lanford, 1988), the human transferrin receptor (Domingo and Trowbridge, 1988), and the major surface glycoprotein of extracellular AcMNPV particles (Roberts and Faulkner, 1989), all of which were palmitylated. The hepatitis B virus surface antigen (HBsAg) (Lanford et al., 1989) is one of several proteins known to undergo myristylation in Sf cells.

Thus far, the only protein processing pathway thought to be missing in Sf cells is one which culminates with the α-amidation of a carboxy terminal amino acid, which was not detected when the human gastrin releasing peptide was expressed in Sf-9 cells (LeBacq-Verheyden et al., 1988). Gastrin releasing peptide is normally synthesized in neuroendocrine tissues as a precursor with a carboxy terminal prosequence. The prosequence is subsequently cleaved, two more amino acids are removed, and an α-amidation reaction occurs, which results in the transfer of an amide group from glycine to methionine. The internal cleavage of the gastrin releasing peptide prosequence occurred in Sf-9 cells, but the removal of the additional carboxy terminal amino acids and/or the amidation reaction itself did not.

Protein-Protein Interactions

The supramolecular assembly of foreign proteins expressed in the baculovirus system, including the formation of both homocomplexes and heterocomplexes, has been examined in several different studies. Human transferrin receptor formed homodimers through interchain disulfide bonds, but only inefficiently, as 30% of the product remained as a monomer (Domingo and Trowbridge, 1988). In contrast, v-sis/ platelet-derived growth factor dimerized efficiently, with only small amounts remaining in the monomer form (Giese et al., 1989). Larger homooligomeric particles were observed upon expression of HBsAg (Kang et al., 1987; Lanford et al., 1989), the major capsid antigen of simian rotavirus (VP6; Estes et al., 1987), a nonstructural protein of bluetongue virus (NS1; Urakawa and Roy, 1988), and SVT ag (Lanford, 1988). Heterooligomeric particles can be produced when two different antigens are simultaneously expressed in insect cells, as was shown for the influenza virus polymerase proteins, PB1 and PB2 (St. Angelo et al., 1987), and for SVT ag coexpressed with murine p53 (O'Reilly and Miller, 1988). One of the most profound examples of multiple protein-protein interactions was recently reported by Urakawa and coworkers (1989). They used a recombinant baculovirus to express the entire poliovirus genome in insect cells, showed that the virus-encoded proteases processed the polyprotein to VP0, VP1 and VP3, and found that these products assembled into a stable, noninfectious poliovirus-like particle. It must be recognized that each of the protein-protein interactions described above is specific; problems of insolubility and nonspecific aggregation are not typically encountered in the baculovirus system, even for proteins expressed at very high levels.

Protein Targeting

Another question that arose during the early development of the baculovirus expression system was whether the signals that targeted mammalian proteins to specific subcellular destinations would be recognized in lepidopteran insect cells. Today, we know that they are. Proteins destined for secretion or cell surface expression enter the insect cellular secretory pathway and their signal peptides are cleaved accurately by insect signal peptidase, as mentioned above. Examples of heterologous proteins that are known to be secreted from insect cells include human β interferon (Smith, Summers, and Fraser, 1983), interleukin-2 (Smith et al., 1985), and tissue plasminogen activator (Jarvis and Summers, 1989). Among those shown to be expressed on the insect cell surface are the HA glycoprotein of influenza viruses (Kuroda et al., 1986; Possee, 1986), the human epidermal growth factor receptor (Greenfield et al., 1988), and the Sindbis virus envelope glycoproteins (Jarvis, Oker-Blom, and Summers, 1990). It also has recently been shown that polarized cell surface expression of the influenza HA (Kuroda et al., 1989) and AcMNPV gP64 (Keddie, Aponte, and Volkman, 1989) molecules occurs during in vivo infection of insect epithelial cells. Finally, a large number of foreign nuclear proteins have been expressed in the baculovirus system, including all of the nuclear phosphoproteins discussed above, and each was localized to the insect cell nucleus.

Interestingly, lysosomal proteins represent one class of proteins that probably cannot be appropriately targeted in baculovirus-infected cells. Human glucocerebrosidase (Martin et al., 1988) and β-hexosaminidase B (Boose et al., 1990) were both secreted from Sf-9 cells and, in the latter study, it was specifically shown that the protein lacked mannose-6-phosphate, which is necessary for targeting these proteins to the lysosomes (Kornfeld and Mellman, 1989). However, the targeting signal in lysosomal proteins differs from other targeting signals in that it is generated post translationally.

Summary: Protein Processing in Baculovirus-Infected Insect Cells

Overall, the available evidence strongly supports the general conclusion that foreign proteins are usually appropriately processed and targeted in lepidopteran insect cells. The only clear exceptions to this generalization are the absence of α-amidation and certain proteolytic cleavages at internal recognition sites containing basic amino acids. There also appear to be differences in the N-glycosylation pathway, but this question needs to be reexamined. Finally, lysosomal proteins might not be appropriately targeted in these cells, due to the inability of the cells to synthesize mannose-6-phosphate. Despite this wealth of accumulated information, there are important questions on the protein processing and targeting capabilities of the baculovirus expression system that need to be addressed. For example, the kinetics and efficiencies of most of the these processes are unknown. Subtle differences in the rates and efficiencies of protein processing and targeting might have a major influence upon the overall levels of expression obtained for a particular protein product. Also, it must be remembered that recombinant baculoviruses are essentially intact and are fully capable of carrying out a lytic infection in cultured cells. Thus, it is important that we begin to understand the influence of baculovirus infection on the host cell and, especially, on cellular protein processing pathways.

Previously, it was shown that the secretion of human tissue plasminogen activator (TPA) decreases dramatically with time of recombinant baculovirus infection in Sf-9 cells (Jarvis and Summers, 1989). This suggested that the infection might have an adverse effect upon the host cell secretory pathway. Based upon this idea, we developed a new approach for baculovirus-based foreign gene expression that eliminates this potentially adverse effect by circumventing the need for a viral infection. The approach involves the use of the promoter from an immediate early baculovirus gene, IE1, to produce stably transformed lepidopteran insect lines that can express a foreign gene product continuously.

8.3 INSECT CELL TRANSFORMATION

IE1-Mediated Expression in Lepidopteran Insect Cells

Guarino and Summers (1986a, 1987) found that the promoters from the AcMNPV IE1 (immediate early) and 39K (delayed early) genes could induce transient foreign gene expression in uninfected Sf-9 cells. We designed further experiments to determine if these promoters could direct continuous foreign gene expression in these cells, as well (Jarvis et al., 1990). Plasmids were constructed in which a bacterial neomycin resistance (neo-r) gene was inserted downstream of the IE1 or polyhedrin promoter, or the 39K promotor with and without the hr5 enhancer. Each plasmid was used to transfect Sf-9 cells by a standard calcium phosphate procedure (Summers and Smith, 1987). Cotransfections of the 39K promotor constructs and the IE1 plasmid were also done. The cells were allowed to recover for 24 hr, then subcultured into replicate plates. Half of the cultures were fed with control medium, while the other half were fed with medium containing G418, a neomycin antibiotic. At 4 day intervals, viable cell counts were performed in triplicate to determine the average number of cells surviving in the presence or absence of G418. This served as an indicator of continuous neo-r gene expression (Figure 2). As anticipated, the neo-r gene was not expressed by the polyhedrin promoter (510Neo), which is inactive in the absence of other viral gene products. Conversely, both the IE1 promoter (IE1Neo) and the 39K promoter plus the hr5 enhancer (39E + Neo + IE1) could induce expression of the neo-r gene. With the 39K promoter, large

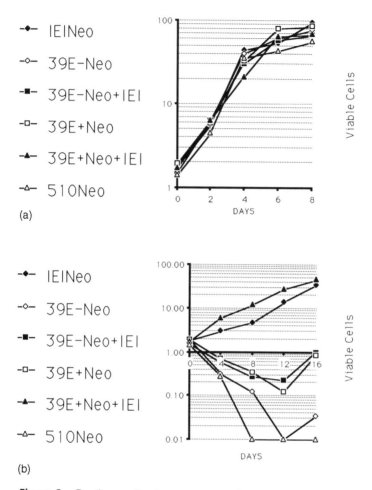

Figure 2 Continuous foreign gene expression with baculovirus promoters. Sf-9 cells were transfected with the plasmids indicated on the left of each figure. These contained the neo resistance gene placed under the control of the AcMNPV IE1 promoter (IE1Neo), the 39K promoter in the absence (39E−Neo) or presence (39E+Neo) of the hr5 enhancer, or the polyhedrin promoter (510Neo). In some cases, plasmids were cotransfected with a second plasmid encoding the AcMNPV IE1 gene product (+ IE1). The cells were incubated for 24 hr after transfection, then subcultured into replicate plates. Half of the plates from each transfection were fed with control medium, while the other half were fed with medium containing G418. At the indicated times post subculture, viable cell counts were performed in triplicate. The average viable cell density in the presence or absence of G418 after transfection with the various plasmids was plotted against days post subculture.

numbers of surviving cells were obtained only if both the hr5 enhancer and the IE1 gene were present. These results agreed well with the results of previous transient expression assays, which had established that the 39K promoter could be activated by hr5, a cis activating enhancer element, and by IE1, a trans activating protein (Guarino and Summers, 1986a, 1986b, 1987; Guarino, Gonzalez, and Summers, 1986). Approximately equal numbers of G418-resistant cells were obtained with either the IE1 promoter alone or with the 39K promoter in combination with hr5 and IE1. These results indicated that, under the right conditions, either of these promoters could be used to transform Sf-9 cells and obtain continuous foreign gene expression.

Subsequent experiments were designed to produce transformed Sf-9 cell clones that would continuously express the *E. coli* β-galactosidase (β-gal) gene or the human TPA gene under the control of the IE1 promoter (Jarvis et al., 1990). Plasmids were constructed in which the β-gal or TPA gene was inserted downstream of the IE1 promoter, then Sf-9 cells were cotransfected with the IE1Neo plus IE1β-gal or IE1Neo plus IE1TPA plasmids, and G418-resistant clones were isolated and amplified. Immunoprecipitation and western blotting experiments revealed that over 50% of the G418-resistant clones also expressed the second gene product, either β-gal or TPA. Molecular analysis of several positive clones revealed that they contained integrated plasmid sequences and, in extended studies of one β-gal clone, it was found that the integrated sequences were maintained for over 50 passages (about 6 months) in culture. In addition, S1 nuclease analyses showed that transcription of the integrated sequences initiated specifically within the IE1 promoter. The amount of β-gal produced by the transformed Sf-9 cells was roughly 1000-fold lower than the amount produced by Sf-9 cells infected with a recombinant baculovirus, where β-gal is very highly expressed (about 200 mg/liter). However, the transformants produced β-gal continuously for over 100 passages in culture (about 1 year) and they produced a more homogeneous β-gal product. Human TPA, a secreted glycoprotein that is relatively poorly expressed in the baculovirus system ($<$ 1 mg/lit), was expressed at comparable levels in transformed Sf-9 cells. Furthermore, the transformed cells secreted virtually all of the newly synthesized TPA product. In contrast, under steady state conditions approximately two thirds of the TPA produced in baculovirus-infected Sf-9 cells is intracellular and significant amounts remain in the intracellular fraction even after 8 hours of chase time in a pulse-chase experiment. A comparison of the rates of TPA secretion in transformed or infected cells revealed that the transformants secreted TPA about twice as quickly. Together, these results support the idea (Jarvis and Summers, 1989) that baculovirus infection has an adverse effect on the function of the Sf-9 cellular secretory pathway.

Thus, by using the IE1 promoter in conjunction with a selectable marker, the standard cotransfection and antibiotic selection techniques developed in mammalian cell systems (Wigler et al., 1980; Southern and Berg, 1982) have been applied for the isolation of stably transformed Sf-9 clones that will express a second gene product continuously. When comparing the relative advantages and disadvantages of expresssing a foreign gene product in stably transformed or baculovirus-infected lepidopteran insect cells, there are several important points to consider. The first is that baculovirus-infected cells are renowned for the ability to provide extremely high levels of foreign gene expression, based upon the strength of the polyhedrin (or p10) promoter. The transformed cell approach, in which transcription is controlled by the relatively weak IE1 promoter, does not provide equally high levels of expression. This is an obvious disadvantage and, if transformed Sf-9 cells are to become widely used for foreign gene expression, the levels of expression should be improved. The second point, however, is that there are some recombinant gene products, like human TPA, that are not highly expressed in baculovirus-infected cells. In this case, where approximately equivalent levels of expression can be achieved with transformed cells, the ability of the cells to continuously express the product becomes a strong advantage. Furthermore, the possibility that a complex recombinant product will be processed faster and more efficiently is highly attractive, although this interpretation is currently based upon results obtained for only one protein. Clearly, the relative protein processing capabilities of infected and transformed Sf-9 cells should be examined further by additional studies of other recombinant proteins. The technical aspects of the two systems of foreign gene expression in lepidopteran insect cells are closely comparable and both are relatively simple. Approximately one month is required for the isolation of working stocks of a recombinant baculovirus or a transformed Sf-9 clone. Finally, the information and experience that has been gained in previous studies on the large-scale cultivation of lepidopteran insect cells (Maiorella et al., 1988; Murhammer and Goochee, 1988) should be equally useful for either method of foreign gene expression in Sf-9 cells.

Drosophila Promoter-Mediated Expression in Dipteran Insect Cells

There are a number of reports on foreign gene expression and transformation of cell lines derived from dipteran insects, including the fruitfly and mosquito. The cell lines commonly used in these studies are Kc (Echalier and Ohanessian, 1970) and Schneider 2 (Schneider, 1972), derived from *Drosophila melanogaster,* and C7, derived from *Aedes albopictus* (Sarver and Stollar, 1977). Generally, expression of a selectable marker and, in some cases, a second gene product, has been accomplished in these cells using constructs containing the promoters from *Drosophila melanogaster* genes.

Bourouis and Jarry (1983) isolated methotrexate resistant transformants of the *Drosophila* Kc cell line. The plasmid used in this study contained a procaryotic dihydrofolate reductase (DHFR) gene cloned downstream of the promoter from the 5' long terminal repeat (LTR) of a *Drosophila* copia element (Flavell et al., 1981). Methotrexate-resistant colonies were obtained after about three weeks of selection and the cells contained integrated plasmid sequences at an estimated level of over 1000 copies per cell. The plasmid sequences were stably integrated, as indicated by the fact that they were maintained for more than three months in culture. The methotrexate-resistant phenotype of the cells indicated that the DHFR gene was expressed, and the presence of DHFR-specific RNA in the cells was documented by northern blotting analysis.

Subsequently, Rio and Rubin (1985) used the cotransformation approach with Schneider 2 cells to produce G418-resistant transformants that could express a second gene. In this study, the copia promoter was used to express a bacterial neomycin resistance gene and the promoter from a *Drosophila* heat shock gene (hsp 70; Pelham, 1982) was used to express the *Drosophila* P transposable element (O'Hare and Rubin, 1983). The transformed cells contained about 10–20 copies of integrated plasmid DNA per cell and these sequences were stable in the absence of neomycin. Expression of the P element sequences was inducible by heat shock, and transcription initiated specifically within the hsp 70 promoter.

The most recent work on stable transformation and foreign gene expression with *Drosophila* cell lines has been carried out by Rosenberg and coworkers (van der Straten et al., 1989a, 1989b; Johansen et al., 1989). In one study, these investigators compared the quality of a number of different antibiotics, including neomycin, hygromycin, and methotrexate, for use as selectable markers in this system (van der Straten et al., 1989a). The genes imparting resistance to these drugs were cloned under the control of the copia 5' LTR and used to select transformed Schneider cells. The results of this survey revealed that hygromycin was the drug of choice. Larger numbers of spontaneous transformants were obtained with G418, and the selection procedure was much more time-consuming with methotrexate. Subsequently, transformants were produced that contained the *E. coli* galactokinase gene under the control of a number of different promoters, including the copia LTR, the *Drosophila* metallothionein promoter (Lastowsky-Perry, Otto, and Maroni, 1985), or the SV40 early promoter. The SV40 early promoter was only weakly active in Schneider cells, while the copia LTR and the metallothionein promoter provided constitutive or inducible foreign gene expression, respectively. The levels of foreign gene expression obtained with these promoters was shown to be directly dependent upon the number of integrated copies of the foreign gene. The highest levels of expression were about 2–3 ng of galactokinase per ug of total protein. The metallothionein promoter was tightly regulated and expression could be induced by about 50-fold with the addition of cadmium. This proved to be especially useful for the expression of the product of a *ras* oncogene, which was toxic when expressed constitutively in insect cells (van der Straten et al., 1989a; Johansen et al., 1989). Healthy transformants were isolated in the absence of inducer, then expression was induced with cadmium and the p21 ras protein, which constituted about 0.2–0.5% of the total cellular protein, was isolated. Importantly, these authors found that the metallothionein promoter remained inducible even after the transformants had been maintained in culture for more than ten months.

While work on stable transformation and foreign gene expression has been quite successful in *Drosophila* cell systems, technical difficulties with transfection procedures and the relative instability of transformants have hampered analogous work in mosquito cell systems. Durbin and Fallon (1985) reported transient expression of a bacterial chloramphenicol acetyltransferase (CAT) gene under the control of a *Drosophila* heat shock promoter (hsp 70) in *Aedes albopictus* (C7) cells. CAT expression was inducible in this system. A key contribution of this study was the finding that the efficiency of transfection in these mosquito cells could be substantially increased by using polybrene (Kawai and Nishizawa, 1984) instead of calcium phosphate (Wigler et al., 1980). Stable transformants were not selected in this study, but it was reported that plasmid sequences were maintained in the transfected cells for at least four days.

Stable transformation of mosquito cells has been recently accomplished. In one study, it was shown that the *Drosophila* hsp70 promoter could be used to express different selectable markers in *Aedes albopictus* cells and stably transformed clones could be isolated using this approach (Monroe et al., 1989). Interestingly, it was found that the hsp 70 promoter did not always behave as an inducible promoter when incorporated into the genome of these mosquito cells. Another recent study has shown that the *Drosophila* P transposable element could be used, in conjunction with G418 selection, for the stable transformation of an *Aedes aegypti* mosquito cell line (Lycett, Eggleston, and Crampton, 1989).

CONCLUDING REMARKS

The purpose of this chapter was to review the current state of the art in foreign gene expression in insect cell systems. The baculovirus system is the most widely used approach and, over the past five years, a tremendous amount has been learned about the characteristics and capabilities of this system. Simultaneously, we have accumulated a wealth of information on the basic cell biology of lepidoptera and we are beginning to gather information on the baculovirus-host cell interaction. This has led to the emergence of the stable transformation approach, which can provide insect cell lines capable of expressing a foreign gene product continuously in the absence of a viral infection. This approach is currently being developed in both lepidopteran and dipteran insect cell systems. With further work in this area, it is possible that the transformed insect cell approach will become as useful for foreign gene expression as the baculovirus approach. In any event, it will be extremely gratifying if this work contributes as much new knowledge of insect cell systems in the next five years as has been gained from work on baculovirus-mediated foreign gene expression over the past five years.

REFERENCES

Blissard, G. W. and Rohrmann, G. F. (1990). Baculovirus diversity and molecular biology, *Ann. Rev. Entomol., 35:*127–155.

Boose, J. A., Tifft, C. J., Proia, R. L., and Meyerowitz, R. (1990). Synthesis of a human lysosomal enzyme, β-hexosaminidase B, using the baculovirus expression system, *Prot. Exp. Pur., 1:*111–120.

Bourouis, M. and Jarry, B. (1983). Vectors containing a procaryotic dihydrofolate reductase gene transform Drosophila cells to methotrexate-resistance, *EMBO J., 2:*1099–1104.

Butters, T. D. and Hughes, R. C. (1981). Isolation and characterization of mosquito cell membrane glycoproteins, *Biochim. Biophys. Acta, 640:*655–671.

Butters, T. D., Hughes, R. C., and Vischer, P. (1981). Steps in the biosynthesis of mosquito cell membrane glycoproteins and the effects of tunicamycin, *Biochim. Biophys. Acta, 640:*672–686.

Davidson, D. J., Fraser, M. J., and Castellino, F. J. (1990). Oligosaccharide processing in the expression of human plasminogen cDNA by lepidopteran insect (*Spodoptera frugiperda*) cells, *Biochemistry, 29:*5584–5590.

Domingo, D. L. and Trowbridge, I. S. (1988). Characterization of the human transferrin receptor produced in a baculovirus expression system, *J. Biol. Chem., 263:*13386–13392.

Durbin, J. E. and Fallon, A. M. (1985). Transient expression of the chloramphenicol acetyltransferase gene in cultured mosquito cells, *Gene, 36:*173–178.

Echalier, G. and Ohanessian, A. (1970). In vitro culture of *Drosophila melanogaster* cells, *In Vitro, 6:*162–172.

Ellis, L., Levitan, A., Cobb, M. H., and Ramos, P. (1988). Efficient expression in insect cells of a soluble, active human insulin receptor protein-tyrosine kinase domain by use of a baculovirus vector, *J. Virol., 62:*1634–1639.

Emery, V. C. and Bishop, D. H. L. (1987). The development of multiple expression vectors for high level synthesis of eukaryotic proteins: Expression of LCMV-N and AcNPV polyhedrin protein by a recombinant baculovirus, *Protein Engineering, 1:*359–366.

Estes, M. K., Crawford, S. E., Penaranda, M. E., Petrie, B. L., Burns, J. W., Chan, W.-K., Ericson, B., Smith, G. E., and Summers, M. D. (1987). Synthesis and immunogenicity of the rotavirus major capsid antigen using a baculovirus expression system, *J. Virol., 61:*1488–1494.

Flavell, A. J., Levis, R., Simon, M., and Rubin, G. M. (1981). The 5′ termini of RNAs encoded by the transposable element copia, *Nucleic Acids Res., 9:*6279–6291.

Friesen, P. D. and Miller, L. K. (1986). The regulation of baculovirus gene expression, *Curr. Top. Microbiol. Immunol., 131:*31–49.

Giese, N., May-Siroff, M., LaRochelle, W. J., Van Wyke Coelingh, K., and Aaronson, S. A. (1989). Expression and purification of biologically active v-*sis*/platelet-derived growth factor B protein by using a baculovirus vector system, *J. Virol., 63:*3080–3086.

Grabowski, G. A., White, W. R., and Grace, M. E. (1989). Expression of functional human acid β-glucosidase in COS-1 and *Spodoptera frugiperda* cells, *Enzyme*, in press.

Granados, R. R. and Federici, B. A., (eds.), (1986). *The Biology of Baculoviruses*, CRC Press, Boca Raton, Florida.

Greenfield, C., Patel, G., Clark, S., Jones, N., and Waterfield, M. D. (1988). Expression of the human EGF receptor with ligand-stimulatable kinase activity in insect cells using a baculovirus vector, *EMBO J., 7:*139–146.

Guarino, L. A. and Summers, M. D. (1986a). Functional mapping of a transactivating gene required for expression of a baculovirus delayed-early gene, *J. Virol., 57:*563–571.

Guarino, L. A. and Summers, M. D. (1986b). Interspersed homologous DNA of *Autographa californica* nuclear polyhedrosis virus enhances delayed-early gene expression, *J. Virol., 60:*215–223.

Guarino, L. A., Gonzalez, M. A., and Summers, M. D. (1986). Complete sequence and enhancer function of the homologous DNA regions of *Autographa californica* nuclear polyhedrosis virus, *J. Virol., 60:*224–229.

Guarino, L. A. and Summers, M. D. (1987). Nucleotide sequence and temporal expression of a baculovirus regulatory gene, *J. Virol., 61:*2091–2099.

Guarino, L. A. (1989). Enhancers of early gene expression, *Invertebrate Cell System Applications Volume I* (J. Mitsuhashi, ed.), CRC Press, Boca Raton, Florida, pp. 211–219.

Hanover, J. A. and Lennarz, W. J. (1981). Transmembrane assembly of membrane and secretory glycoproteins, *Arch. Biochem. Biophys., 211:*1–19.

Hart, G. W., Haltiwanger, R. S., Holt, G. D., and Kelly, W. G. (1989). Glycosylation in the nucleus and cytoplasm, *Ann. Rev. Biochem., 58:*841–874.

Hink, W. F. and Vail, P. V. (1973). A plaque assay for titration of alfalfa looper nuclear polyhedrosis virus in a cabbage looper (TN-368) cell line, *J. Invert. Path., 22:*168–174.

Hink, W. F., Thomsen, D. R., Meyer, A. L., Davidson, D. D., and Castellino, F. J. (1991). Expression of three recombinant proteins using baculovirus vectors in 23 insect cell lines, *Biotech. Progress, 7:*9–14.

Hooft van Iddekinge, B. J. L., Smith, G. E., and Summers, M. D. (1983). Nucleotide sequence of the polyhedrin gene of *Autographa californica* nuclear polyhedrosis virus, *Virology, 131:*561–565.

Hoss, A., Moarefi, I., Scheidtmann, K.-H., Cisek, L., Corden, J. L., Dornreiter, I., Arthur, A. K., and Fanning. E. (1990). Altered phosphorylation of simian virus 40 large T antigen expressed in insect cells by using a baculovirus vector, *J. Virol., 64:*4799–4807.

Hsieh, P. and Robbins, P. W. (1984). Regulation of asparagine-linked oligosaccharide processing. Oligosaccharide processing in aedes albopictus mosquito cells, *J. Biol. Chem., 259:*2375–2382.

Jarvis, D. L. and M. D. Summers. (1989). Glycosylation and secretion of human tissue plasminogen activator in recombinant baculovirus-infected insect cells, *Mol. Cell. Biol., 9:*214–223.

Jarvis, D. L., Oker-Blom, C., and Summers, M. D. (1990). The role of glycosylation in the transport of foreign glycoproteins through the secretory pathway of Lepidopteran insect cells, *J. Cell. Biochem., 42:*181–191.

Jarvis, D. L., Fleming, J. G. W., Kovacs, G. R., Summers, M. D., and Guarino, L. A. (1990). Use of early baculovirus promoters for continuous expression and efficient processing of foreign gene products in stably transformed lepidopteran insect cells. *Bio/Technology,* in press.

Jeang, K.-T., Giam, C.-Z., Nerenberg, M., and Khoury, G. (1987). Abundant synthesis of functional human T cell leukemia virus type I p40^X protein in eucaryotic cells by using a baculovirus expression vector, *J. Virol., 61:*708–713.

Johansen, H., Van der Straten, A., Sweet, R, Otto, E., Maroni, G., and Rosenberg, M. (1989). Regulated expression at high copy number allows production of a growth-inhibitory oncogene product in *Drosophila* Schneider cells, *Genes and Dev., 3:*882–889.

Kang, C. Y., Bishop, D. H. L., Seo, J.-S., Matsuura, Y., and Choe, M. (1987). Secretion of particles of hepatitis B virus surface antigen from insect cells using a baculovirus vector, *J. Gen Virol., 68:*2607–2613.

Kang, C. Y. (1988). Baculovirus vectors for expression of foreign genes, *Adv. Virus Res., 35:*177–192.

Kawai, S. and Nishizawa, M. (1984). New procedure for DNA transfection with polycation and dimethylsulfoxide, *Mol. Cell. Biol., 4:*1172–1174.

Keddie, B. A., Aponte, G. W., and Volkman, L. E. (1989). The pathway of infection of *Autographa californica* nuclear polyhedrosis virus in an insect host, *Science, 243:*1728–1730.

Kornfeld, S. and Mellman, I. (1989). The biogenesis of lysosomes, *Ann. Rev. Cell Biol., 5:*483–525.

Kuroda, K., Hauser, C., Rott, R., Klenk, H.-D., and Doerfler, W. (1986). Expression of the influenza virus haemagglutinin in insect cells by a baculovirus vector, *EMBO J, 5:*1359–1365.

Kuroda, K., Groner, A., Frese, K., Drenckhahn, D., Hauser, C., Rott, R., Doerfler, W., and Klenk, H.-D. (1989). Synthesis of biologically active influenza virus hemagglutinin in insect larvae, *J. Virol., 63:*1677–1685.

Kuroda, K., Geyer, H., Geyer, R., Doerfler, W., and Klenk, H.-D. (1990). The oligosaccharides of influenza virus hemagglutinin expressed in insect cells by a baculovirus vector, *Virology, 174:*418–429.

Kuzio, J., Rohel, D. Z., Curry, C. J., Krebs, A., Carstens, E. B., and Faulkner, P. (1984). Nucleotide sequence of the p10 polypeptide gene of *Autographa californica* nuclear polyhedrosis virus, *Virology, 139:*414–418.

Lanford, R. E. (1988). Expression of simian virus 40 T antigen in insect cells using a baculovirus expression vector, *Virology, 167:*72–81.

Lanford, R. E., Luckow, V., Kennedy, R. C., Dreesman, G. R., Notvall, L., and Summers, M. D. (1989). Expression and characterization of hepatitis B virus surface antigen polypeptides in insect cells with a baculovirus expression system, *J. Virol., 63:*1549–1557.

Lastowsky-Perry, D., Otto, E., and Maroni, G. (1985). Nucleotide sequence and expression of a *Drosophila* metallothionein, *J. Biol. Chem., 260:*1527–1530.

LeBacq-Verheyden, A.-M., Kasprzyk, P. G., Raum, M. G., van Wyke Coelingh, K., LeBacQ, J. A., and Battey, J. F. (1988). Posttranslational processing of endogenous and of baculovirus-expressed human gastrin-releasing peptide precursor, *Mol. Cell. Biol., 8:*3129–3135.

Luckow, V. L. and Summers, M. D. (1988a). Trends in the development of baculovirus expression vectors, *Bio/Technology, 6:*47–55.

Luckow, V. A. and Summers, M. D. (1988b). Signals important for high-level expression of foreign genes in *Autographa californica* nuclear polyhedrosis virus expression vectors, *Virology, 167:*56–71.

Luckow, V. A. and Summers, M. D. (1989). High level expression of non-fused foreign genes with *Autographa californica* nuclear polyhedrosis virus expression vectors, *Virology, 170:*31–39.

Lycett, G., Eggleston, P., and Crampton, J. (1989). "DNA Transfection of an Aedes aegypti Mosquito Cell Line," Abstracts of the International Symposium on Molecular Insect Sciences, p. 66.

Maiorella, B., Inlow, D., Shauger, A., and Harano, D. (1988). Large-scale insect cell culture for recombinant protein production, *Bio/Technology, 6:*1406–1410.

Martin, B. M., Tsuji, S., LaMarca, M. E., Maysak, M., Eliason, W., and Ginns, E. I. (1988). Glycosylation and processing of high levels of active human glucocerebrosidase in invertebrate cells using a baculovirus expression vector, *DNA, 7:*99–106.

Matsuura, Y., Possee, R. D., and Bishop, D. H. L. (1987). Baculovirus expression vectors: The requirements for high level expression of proteins, including glycoproteins, *J. Gen. Virol., 68:*1233–1250.

Miller, L. K. (1988). Baculoviruses as gene expression vectors, *Ann. Rev. Microbiol., 42:*177–199.

Monroe, T. J., Carlson, J. O., Clemens, D. L., and Beaty, B. J. (1989). "Selectable Markers for Transformation of Mosquito and Mammalian Cells," Abstracts of the International Symposium on Molecular Insect Sciences, p. 72.

Murhammer, D. W. and Goochee, C. F. (1988). Scale-up of insect cell cultures: Protective effects of pluronic F-68, *Bio/Technology, 6:*1411–1418.

Nyunoya, H., Akagi, T., Ogura, T., Maeda, S., and Shimotohno, K. (1988). Evidence for phosphorylation of *trans*-activator p40^X of human T cell leukemia virus type I produced in insect cells with a baculovirus expression vector, *Virology, 167:*538–544.

O'Hare, K. and Rubin, G. M. (1983). Structures of P transposable elements and their sites of insertion and excision in the *Drosophila melanogaster* genome, *Cell, 34:*25–35.

Oker-Blom, C. and Summer, M. D. (1989). Expression of Sindbis virus 26S cDNA in *Spodoptera frugiperda* cells using a baculovirus expression vector, *J. Virol., 63:*1256–1264.

Oker-Blom, C., Pettersson, R. F., and Summers, M. D. (1989). Baculovirus polyhedrion promoter directed expression of rubella virus envelope glycoproteins, E1 and E2, in *Spodoptera frugiperda* cells, *Virology, 172:*82–91.

O'Reilly, D. R. and Miller, L. K. (1988). Expression and complex formation of simian virus 40 large T antigen and mouse p53 in insect cells, *J. Virol., 62:*3109–3119.

Pelham, H. R. B. 1982. A regulatory upstream promoter element in the Drosophila hsp70 heat shock gene, *Cell, 30:*517–528.

Possee, R. D. (1986). Cell-surface expression of influenza virus hemagglutinin in insect cells using a baculovirus vector, *Virus Res., 5:*43–59.

Possee, R. D. and Howard, S. C. (1987). Analysis of the polyhedrin gene promoter of the *Autographa californica* nuclear polyhedrosis virus, *Nuc. Acids Res., 15:*10233–10248.

Rankin, C., Ooi, B. R., and Miller, L. K. (1988). Eight base pairs encompassing the transcriptional start point are the major determinant for baculovirus polyhedrin gene expression, *Gene, 70:*39–49.

Rio, D. C. and Rubin, G. M. (1985). Transformation of cultured *Drosophila melanogaster* cells with a dominant selectable marker, *Mol. Cell. Biol., 5:*1833–1838.

Roberts, T. E. and Faulkner, P. (1989). Fatty acid acylation of the 67K envelope glycoprotein of a baculovirus: *Autographa californica* nuclear polyhedrosis virus, *Virology, 172:*377–381.

Rusche, J. R., Lynn, D. L., Robert-Guroff, M., Langlois, A. J., Lyerly, H. K., Carson, H., Krohn, K., Ranki, A., Gallo, R. C., Bolognesi, D. P., Putney, S. D., and Matthews, T. J. (1987). Humoral immune response to the entire human immunodeficiency virus envelope glycoprotein made in insect cells, *Proc. Natl. Acad. Sci. U.S.A., 84:*6924–6928.

Sarver, N. and Stollar, V. (1977). Sindbis virus-induced cytopathic effect in clones of *Aedes albopictus* (Singh) cells, *Virology, 80:*390–400.

Schneider, I. (1972). Cell lines derived from late embryonic stages of *Drosophila melanogaster, J. Embryol. Exp. Morphol. 27:*353–356.

Sissom, J. and Ellis, L. (1989). Secretion of the extracellular domain of the human insulin receptor from insect cells by use of a baculovirus vector, *Biochem. J., 261:*119–126.

Smith, G. E., Vlak, J. M., and Summers, M. D. (1983). Physical analysis of *Autographa californica* nuclear polyhedrosis virus transcripts for polyhedrin and a 10,000-molecular weight protein, *J. Virol., 45:*215–225.

Smith, G. E., Fraser. M. J., and Summers, M. D. (1983). Molecular engineering of the *Autographa californica* nuclear polyhedrosis virus genome: Deletion mutations within the polyhedrin gene, *J. Virol., 46:*584–593.

Smith, G. E., Summers, M. D., and Fraser, M. J. (1983). Production of human beta interferon in insect cells infected with a baculovirus expression vector, *Mol. Cell. Biol., 3:*2156–2165.

Smith, G. E., Ju, G., Ericson, B. L., Moschera, J., Lahm, H.-W., Chizzonite, R., and Summers, M. D. (1985). Modification and secretion of human interleukin 2 produced in insect cells by a baculovirus expression vector, *Proc. Natl. Acad. Sci. U.S.A., 82:*8404–8408.

Southern, P. J. and Berg, P. (1982). Transformation of mammalian cells to antibiotic resistance with a bacterial gene under control of the SV40 early region promoter, *J. Mol. Appl. Gen., 1:*327–341.

St. Angelo, C., Smith, G. E., Summers, M. D., and Krug, R. M. (1987). Two of the three influenza viral polymerase proteins expressed by using baculovirus vectors form a complex in insect cells, *J. Virol., 61:*361–365.

Summers, M. D. and Smith, G. E. (1987). *A manual of methods for baculovirus vectors and insect cell culture procedures.* Texas Agricultural Experiment Station Bulletin No. 1555.

Thomsen, D. R., Post, L. E., and Elhammer, A. P. (1990). Structure of O-glycosidically linked oligosaccharides synthesized by the insect cell line Sf-9, *J. Cell. Biochem., 43:*67–79.

Urakawa, T. and Roy, P. (1988). Bluetongue virus tubules made in insect cells by recombinant baculoviruses: Expression of the NS1 gene of bluetongue virus serotype 10, *J. Virol., 62:*3919–3927.

Urakawa, T., Ferguson, M., Minor, P. D., Cooper, J., Sullivan, M., Almond, J. W., and Bishop, D. H. L. (1989). Synthesis of immunogenic, but non-infectious, poliovirus particles in insect cells by a baculovirus expression vector, *J. Gen. Virol., 70:*1453–1463.

Van der Straten, A., Johansen, H., Sweet, R., and Rosenberg, M. (1989a). Efficient expression of foreign genes in cultured *Drosophila melanogaster* cells using hygromycin B selection, *Invertebrate Cell System Applications Volume I* (J. Mitsuhashi, ed.), CRC Press, Boca Raton, Florida, p. 183–195.

Van der Straten, A., Johansen, H., Rosenberg, M., and Sweet, R. (1989b). Novel hygromycin B selection system for the overexpression of a heterologous gene in *Drosophila melanogaster* cultured cells, *Curr. Methods Mol. Biol., 1:*1–8.

Vaughn, J. L., Goodwin, R. H., Thompkins, G. J., and McCawley, P. (1977). The establishment of two insect cell lines from the insect *Spodoptera frugiperda* (Lepidoptera:Noctuidae), *In Vitro, 13:*213–217.

Vlak, J. M., Klinkenberg, F. A., Zaal, K. J. M., Usmany, M., Klinge-Roode, E. C., Geervliet, J. B. F., Roosien, J., and van Lent, J. W. M. (1988). Functional studies on the p10 gene of *Autographa californica* nuclear polyhedrosis virus using a recombinant expressing a p10-β-galactosidase fusion gene, *J. Gen. Virol., 69:*765–776.

Vlak, J. M. and Keus, R. J. A. (1990). Baculovirus expression vector system for production of viral vaccines, *Adv. Biotech. Processes, 14:*91–128.

Volkman, L. E. and Summers, M. D. (1975). Nuclear polyhedrosis virus detection: relative capabilities of clones developed from *Trichoplusia ni* ovarian cell line TN-368 to serve as indicator cells in a plaque assay, *J. Virol., 16:*1630–1637.

Weiss, S. A., Kalter, S. S., Vaughn, J. L., and Dougherty, E. (1980). Effect of nutritional, biological, and biophysical parameters on insect cell culture of large scale production, *In Vitro, 16:*222–223.

Weiss, S. A., Orr, T., Smith, G. C., Kalter, S. S., Vaughn, J. L. and Dougherty, E. M. (1982). Quantitative measurement of oxygen consumption in insect cell culture infected with polyhedrosis virus, *Biotechnol. Bioeng., 24:*1145–1154.

Whitford, M. and Faulkner, P. (1990). "A 42-kDa Structural Polypeptide from the Baculovirus AcMNPV Contains N-acetylglucosamine Linked by O-glycosylation," Abstracts of the Eighth International Congress of Virology, p. 157.

Wigler, M., Perucho, M., Kurtz, D., Dana, S., Pellicer, A., Axel, R., and Silverstein, S. (1980). Transformation of mammalian cells with an amplifiable dominant-acting gene, *Proc. Natl. Acad. Sci. U.S.A., 77:*3567–3570.

Wojchowski, D. M., Orkin, S. H., and Sytkowski, A. J. (1987). Active human erythropoeitin expressed in insect cells using a baculovirus vector: A role for N-linked oligosaccharide, *Biochim. Biophys. Acta, 910:*224–232.

9

The Future of Insect Cell Culture Engineering

Spiros N. Agathos   *Rutgers, The State University of New Jersey, Piscataway, New Jersey*

9.1 INTRODUCTION

Insect cell culture is becoming recognized as an important enabling technology for the production of biologicals, including recombinant proteins and biopesticides, primarily through the use of baculovirus expression vectors (BEV). In an extensive recent review, Luckow (1991) tracks the burgeoning literature in the field of cloning and expression of heterologous genes in lepidopteran insect cells with BEV, and has noted an excess of 350 publications in this area from 1983 till mid 1990, the majority appearing since 1987. The reasons behind this steadily increasing flow of basic and technological information lie in the unique and highly attractive features of this expression system. Among the chief advantages of insect cell-BEV systems are: (1) the very high levels of foreign gene expression compared with most mammalian and several microbial cell systems, (2) the capacity to produce proteins that are correctly processed posttranslationally in the insect host cell, and are, as a result, enzymatically, immunologically, and functionally similar to their authentic counterparts, in contrast with many proteins expressed in bacterial or yeast systems, (3) the helper-independence and inability to activate dormant oncogenes, unlike several mammalian expression systems, and (4) the easy manipulation of BEV, which allows the straightforward insertion of an astonishing variety of coding sequences from practically any cellular or tissue origin, including cytotoxic gene products. These biological features, coupled with the similarities between in vitro insect cell propagation and mammalian cell culture, suggest that the insect cell-BEV system will see an expansion of its use not only for production of biological materials useful in research and product development, but also for bulk manufacturing of therapeutic pharmaceuticals, diagnostics, and biological control agents for agriculture, assuming of course the regulatory hurdles will not be excessive.

It is clear that the lepidopteran cell-BEV system is by far the main driving force behind insect cell culture engineering. In this dominant system, the standard cell line in most widespread use is line Sf-9 from *Spodoptera frugiperda,* and the most common vector for protein expression consists of genetically engineered *Autographa californica* nuclear polyhedrosis virus (AcNPV) (Luckow and Summers, 1988; Miller, 1988; Bishop, 1989a). The cultivation of dipteran cell lines is currently pursued much less than that of lepidopteran cell lines, since the former are not efficient hosts to BEV. Nonetheless, emerging advances in dipteran cell use for regulated protein expression, as illustrated by the Schneider 2 cell line of *Drosophila melanogaster* stably integrating multiple copies of a foreign gene in the insect chromosome, promise to complement the already high potential of the lepidopteran cell-BEV system (Culp et al., 1991).

The future prospects of insect cell culture as a mainstream biotechnological production-scale methodology are shaped by a number of factors, including: (1) progress in basic understanding of the biological functions of insect cells as vehicles for protein synthesis, (2) improvements in the design and operation of appropriate bioreactors, and (3) development of inexpensive and convenient culture media, especially serum-free and low protein media of chemically defined composition. Trends in these areas are briefly outlined below.

9.2 PROGRESS IN INSECT CELL LINES AND EXPRESSION VECTORS

Despite the prodigious number and great variety of proteins that have been cloned and expressed in insect cell culture using BEV (for an updated compilation, numbering over 130 different proteins, see Luckow, 1991), their overwhelming majority has been produced in Sf-9 cells infected with recombinant AcNPV. However, there are alternative insect cell lines and alternative expression vectors suitable for cloning heterologous genes. Among the dozens of established insect cell lines and strains available for in vitro cultivation (Hink and Hall, 1989), many are susceptible to AcNPV or other nuclear polyhedrosis viruses that are specific to the particular insect species. For example, a widely studied cell line, Tn-368 from the cabbage looper, *Trichoplusia ni*, is used for propagation of wild type and recombinant AcNPV (Shuler et al., 1990; Ogonah et al., 1991). Wild type AcNPV is known to infect cultured cell lines from *Estigmene acrea* (saltmarsh caterpillar), *Lymantria dispar* (gypsy moth), *Mamestra brassicae* (cabbage worm), *Heliothis virescens* (tobacco budworm) and others (Lynn and Hink, 1980; McIntosh and Ignoffo, 1989; Granados and Hashimoto, 1989). The *Heliothis zea* (cotton bollworm) nuclear polyhedrosis virus, HzNPV (Fraser, 1989) and the *Bombyx mori* (silkworm) nuclear polyhedrosis virus, BmNPV (Maeda, 1989; Iatrou et al., 1989) are used for BEV development and gene cloning in their respective insect cell lines.

There are few comparative studies of gene expression level and of cellular or volumetric protein productivity in diverse cell lines using the exact same BEV construct under standardized conditions. Such studies are a prerequisite in future efforts to increase consistently the level of expression of recombinant genes in BEV systems. Our current quantitative understanding of the influence that cultured insect cells have on protein expression is minimal, and the few reports comparing protein production in parallel among several cell lines are still largely inconclusive. For instance, a decade ago, when the driving force behind insect cell culture technology was the development of ample quantities of viral insecticides, the replication of wild type AcNPV was compared in cultured insect cell lines from *S. frugiperda*, *E. acrea*, *L. dispar*, *M. brassicae*, and *T. ni* (Lynn and Hink, 1980). A line from *M. brassicae*, IZD-MB0503, was found superior in terms of susceptibility to infection, level of polyhedrin occlusion bodies (POB) and rapidity of extracellular virus production. More recent work (McIntosh and Ignoffo, 1989) comparing cellular growth rate, infectivity and productivity of wild type AcNPV among five cell lines from *S. frugiperda*, *T. ni*, *H. virescens*, *Plutella xylostella* (diamond back moth), and *Anticarsa gemmitalis* (soybean caterpillar) suggested that the highest viral titer producers were the *T. ni* and *H. virescens* cell lines, although the *P. xylostella* cell line was superior in POB yield. Given the still uncertain relationship between intact wild type virus infectivity, polyhedrin gene expression and recombinant (polyhedrin replacement) gene expression in the same host, it is not automatically evident that the number of POB produced per cell and the rate of viral replication are reliable quantitative predictors of the level of foreign gene expression expected from a genetically engineered virus infecting the cell line in question. It is easy to see that this underlying uncertainty also precludes rational projections of protein productivity among diverse host cell lines on the basis of wild type virus infection data in these cell lines. In fact, the progression of the infection and the cytopathic effects observed in Sf-9 cells differ between wild type AcNPV infection and infection with a recombinant β-galactosidase-encoding AcNPV (Schopf et al., 1990). In the next several paragraphs I will attempt to review in some detail the limited data from comparative studies of expression of the same protein(s) in several cell lines, in the hope of illustrating not only the clear need to screen many alternative hosts for optimal expression of a given protein, but also some of the pitfalls associated with the execution of such studies.

Very recently, Hink and coworkers (1991) approached this fundamental question of comparative yields and expression characteristics of the same recombinant protein(s) using BEV to infect numerous insect cell lines in small-scale, static suspension culture. These workers chose three different types of genes for insertion into AcNPV and they used the resulting vectors to infect 23 different lepidopteran cell lines or strains. The genes cloned were: (1) a structural gene from a virus encoding Gp50T, an envelope *O*-linked glycoprotein from pseudorabies virus, (2) a bacterial gene encoding the enzyme β-galactosidase from *E. coli,* and (3) a mammalian gene encoding the human proenzyme HPg, which is cleaved to generate plasmin. The 23 host cell lines or strains originated from 12 different species of lepidopteran insects, including the standard Sf-9 cell line from *S. frugiperda.* For HPg, line IZD-MB0503 from *M. brassicae* had the highest yield, and four other cell lines produced higher levels than Sf-9. For Gp50T, four different cell lines were superior to Sf-9, although certain cell lines produced Gp50T with molecular mass about 1000 daltons larger than that from Sf-9 cells, which indicates perhaps increased glycosylation. Lines Sf-9, IZD-MB0503, and BCIRL-PX2-HNV3 (the latter from *P. xylostella*) produced the highest levels of β-galactosidase. The authors concluded that no single cell line emerged as the highest producer for all three proteins. For two of the recombinant proteins, HPg and β-galactosidase, they reported a direct comparative relationship between wild type AcNPV replication and the level of protein expression in line IZD-MB0503; they also indicated that for four of the cell lines (IZD-MB0503 plus the other three lines in the earlier study by Lynn and Hink, 1980), the relative numbers of POB produced were directly related to the levels of expression of two (HPg and β-

galactosidase) of the heterologous genes. In the same work, Hink and coworkers (1991) examined the foreign protein yields from cells cultivated in serum-free and serum-supplemented media, using cell lines Sf-9, Tn-368, EAA and Sf-21AE, the parent line from which strain Sf-9 was derived. They concluded that, in general, the production levels in serum-free medium were equal to or better than those in serum-supplemented medium. An examination of protein yields between parent cell lines and clonal strains derived from them was inconclusive, in that, for some proteins, yields were similar between parent and clone, whereas for others yields differed. A final conclusion from this comparative study was that medium composition had a much more pronounced effect on heterologous protein expression than on susceptibility of the insect cells to wild type virus, which may indicate that, generally speaking, infectivity data with intact AcNPV may *not* be an adequate predictor of cloned gene expression in a given line.

Research conceived along the lines of the work by Hink et al. (1991), summarized above, is clearly timely and of urgent significance in our quest for more efficient and more widely applicable insect cell lines as hosts for foreign gene expression. Nonetheless, the difficulties (and tedium!) of precise comparisons at all stages and levels of such studies cannot be overlooked. For instance, it is known empirically that environmental factors affecting foreign gene expression in cultured insect cells using BEV include, but are not limited to: cell density, cell viability, growth stage, multiplicity of infection (MOI), time post infection of assessing protein yield, nutrition (media ingredients, serum or substitutes), available dissolved oxygen concentration (DO), pH, osmotic pressure, etc. (Volkman and Knudsen, 1986). Thus, ideally, comparative studies among diverse cell lines should not only employ the exact same BEV construct, but also maintain the same standardized protocol of cell cultivation and infection, which would imply constancy of all the above-mentioned environmental factors. Although some of these parameters can be maintained at substantially similar levels between cell lines (e.g., pH, DO, cell density and MOI at the time of infection), others are harder to control at the desirable values or ranges. For instance, different cell lines have different growth characteristics (e.g., doubling time) in the same cultivation medium, and, thus, growth stage, time post infection of evaluating protein yield, and, perhaps more generally, physiological competence for gene expression may be much more elusive when comparing cells growing in different media. In the study by Hink and coworkers (1991) an effort was made to maintain the same medium, pH, osmotic pressure, DO, cell density, growth stage, and MOI for each cell line, and only the foreign gene encoded by the recombinant AcNPV vector differed. However, different cell lines and strains were cultivated in different media, and exhibited different growth characteristics from one another. Only the comparison among four cell lines (targeted at the influence of serum-free versus serum-supplemented media, as described above) involved a single serum-free medium (ISFM) for all four lines, but three different serum containing media (IPL-41, TNM-FH, and GTC-100). On the other hand, Hink et al. (1991) attempted to standardize the procedure for infecting the cell lines, by keeping the cell density constant at 7.5×10^5 cells/ml and in exponential growth phase when exposed to the virus, by maintaining the same volume of virus inoculum for all lines, and by not carrying any spent media over to the infected cell cultures. There were, however, no continuous kinetic data of cell growth, baculovirus infection, and recombinant protein production in the 23 cell lines tested, and the protein yields were reported on a volumetric basis, with no indication of protein productivity per cell (it is known that smaller cells can reach higher cell densities but produce lower protein levels per cell than larger cells). Thus, measuring protein productivities in different media, and reporting yields only at 2 and 3 days post infection rather than over the whole period of the batch until complete cell lysis, make the above comparisons among the various cell lines problematic. Perhaps more importantly for large-scale applications, it is not yet known to what extent data from static cultivation (T flasks or petri dishes) can be extrapolated to agitated suspension culture (spinner flasks and bioreactors).

In a comparative study between Tn-368 and Sf-9 cells grown in spinner flasks in the same medium, GTC 100 (with 10% serum) and infected with the same AcNPV vector encoding β-galactosidase, Ogonah et al. (1991) found that Tn-368 cells produced about 30% more β-galactosidase per cell, and twice as much enzyme per reactor volume, than Sf-9 cells, while peak β-galactosidase concentration was reached in Tn-368 cells one day earlier than in Sf-9 cells (2.5 vs. 3.5 days post infection). Nonetheless, these authors concluded that, because of their comparative ease of cultivation and higher resistance to shear, Sf-9 cells may be more suitable for large-scale production of recombinant proteins than Tn-368 cells.

Another comparative study of recombinant protein production was conducted recently in the same serum-free medium, ExCell 400, with Sf-9 cells and two alternative candidate cell lines, IPLB-LdEIta (LdEIta) from *L. dispar* and IPLB-HvT1 (HvT1) from *H. virescens* (Betenbaugh et al., 1991). Both LdEIta and HvT1 cells were reported to produce more than 40% higher total levels of recombinant β-galactosidase in static monolayer culture than Sf-9 cells after infection with the same AcNPV construct, but only LdEIta had a higher product yield than Sf-9 (by 25%) on an activity per total cell protein basis. When comparing LdEIta with Sf-9 for the production of a different recombinant product, rotaviral protein VP4, total production levels of this protein were almost double the levels of Sf-9, and the VP4 yields (on a per total protein basis) achieved by LdEIta were also higher (by 38%) than those in Sf-9 monolayer cultures. The choice of total cell protein rather than cell number as the basis for calculating the recombinant protein yield was made by Betenbaugh et al. (1991) because the three cell lines differ in cell size. The comparison on this basis was also facilitated by the absence of serum or other proteins in the medium. In suspension culture experiments performed in spinner flasks, the same workers compared Sf-9 and LdEIta cells in terms of growth characteristics and recombinant β-galactosidase production. While maximum viable cell densities for LdEIta were about half the values achieved by Sf-9, the LdEIta cells were much larger than Sf-9 cells, as were their total final protein levels. Additionally, LdEIta cells grew as aggregates of 50–200 cells and at longer doubling times than Sf-9 cells, which grew as single cells or small aggregates. The final β-galactosidase yields on a per total cell protein basis were similar to those observed in the monolayer cultures, i.e., LdEIta cells were more productive (by 38%) than Sf-9 cells, whereas total (volumetric basis) enzyme production in LdEIta cells was double that of Sf-9 cells. An examination of the time courses of recombinant protein production in the two cell lines showed that the pronounced difference in total productivity was partially due to the longer post infection viability and sustained gene expression manifested by the LdEIta cell line compared to Sf-9 (Betenbaugh et al., 1991). Long viability of the cells during the course of the infection may not be as important a factor in the overall productivity of a lepidopteran cell-BEV system, since, for instance, proteolytic degradation of the foreign protein product in the infection stage may adversely influence its yield (Licari and Bailey, 1991).

A cursory assessment of the qualitative and, wherever possible, quantitative trends in β-galactosidase production among various strains that were compared in the three characteristic studies chosen for review here (Hink et al. 1991; Ogonah et al., 1991; Betenbaugh et al. 1991) may leave the reader with contradictory conclusions. As pointed out above, scrupulously standardized conditions of cultivation and infection are crucial in establishing the host cell system of choice for the abundant expression of a given recombinant protein through insect cell culture technology. More fundamentally, a deeper understanding of the various factors having an influence on the final titer, the integrity and function of the foreign protein product, and the relative robustness of each cell line under bioreactor cultivation conditions, will be necessary before being able to rationally select the best host cell system for the expression of a given heterologous protein in insect cells.

In addition to the influence of the host cell line, the construction of the baculovirus vector, i.e., the exact location in the viral genome where the recombinant gene is inserted, is clearly of fundamental significance in achieving high levels of protein expression. For instance, even among similar BEV constructs, differences in the point of insertion by a few base pairs can have a profound effect on the levels of expression, which can differ by as much as 20-fold (Matsuura et al., 1986). Efforts will continue in the area of vector construction for high level expression of foreign genes that are not fused with the polyhedrin gene. Successful strategies in this area include the alteration of downstream sequences (Ooi et al., 1989) or of the polyhedrin ATG start codon to ATT or ATC by site-directed mutagenesis (Luckow and Summers, 1989; Page et al., 1989). Analysis of expressed proteins from genes cloned in-frame with various alterations in the polyhedrin initiator codon will not only result in further enhancement of the translational efficiency of foreign gene expression in the insect cell-BEV system, but may also offer new insights in our current understanding of the rules underlying alternative initiation codons in eukaryotic systems. There is increasing evidence that the secondary structure of the mRNA near the initiation codon influences decisively the initiation of translation at non-AUG codons (Kozak, 1990).

The expression of more than one foreign protein, which is now possible through the construction of multiple gene expression vectors (French and Roy, 1990), will provide further impetus for practical applications. For instance, in work by Takehara and coworkers (1988), both surface and core antigens of the hepatitis B virus were expressed, each under the control of the polyhedrin promoter in opposite orientations. Additionally, the exploration of strong promoters other than the polyhedrin promoter, that drive the expression of other nonessential late genes, may offer new options in the quest for abundant foreign gene expression. One such candidate is the p10 promoter (Weyer and Possee, 1989), which drives the expression of a protein that seems nonessential in the replication and production of extracellular or occluded baculovirus, even though it may have some role in the cytopathic effects (loss of viability and eventual lysis) of the infected insect cells (Williams et al., 1989). Clearly, the use of novel expression vectors under the control of a promoter such as p10 must offer significant advantages over vectors that use the polyhedrin promoter, if they are to be considered for extensive BEV construction.

A new development that may extend the applicability of lepidopteran cell lines beyond the lytic baculovirus-based gene expression, is the stable transformation of insect cells with plasmids containing foreign genes under the control of the promoter of IE1, which is an immediate early baculovirus gene (Jarvis et al., 1990). The complex human glycoprotein, tissue plasminogen activator (TPA), was recently shown to be produced more efficiently in stably transformed Sf-9 cells compared to normally (i.e., lytically) infected cells of the same line (Jarvis et al., 1990). In general, expression levels for secreted, glycosylated proteins using the standard lytic BEV systems in insect cells are relatively low in comparison to other recombinant products (Luckow, 1991). Thus, the optimal expression of complex, posttranslationally modified proteins might be one area where the stable transfection of insect cells with genetically engineered vectors under the control of early promoters could emerge as complementary to the standard lytic BEV approach.

9.3 BIOREACTORS FOR INSECT CELL CULTURE

Since insect cell cultivation is, in essence, a variant of mammalian cell culture, the process options for commercial-scale insect cell propagation, in terms of hardware and operational mode, revolve around the same basic repertoire of bioreactor designs and functional strategies as those developed or under development for mammalian cell-based bioprocesses. Clearly, despite the numerous available reactor designs, including an abundance of materials for cell immobilization, the operational alternatives in mode of bioprocessing are limited. Cells are grown either in batch or (semi-) continuously, in suspension or attached form. It can be easily seen from the literature of the last decade, which illustrates the decisive advancement of the art and science of animal cell cultivation for industrial applications, that among the many alternative bioreactor designs (Griffiths, 1990; Hu and Peshwa, 1991) no single system has proven to be universally superior in productivity and scalability. Indeed, a good case can be made that, with an adequate knowledge of the physiological requirements of the subject animal cell line, one can grow the cells in such simple equipment as shake flasks, large spinner flasks, and minimally modified microbial fermentors, with the same yields and productivities as those obtainable from most of the specialized animal cell reactor systems on the market today (Persson et al, 1991). Therefore, trends in efficient bioreactor cultivation of insect cells most likely will be shaped by the evolving technology in other animal cell propagation. In terms of bioprocessing implications, it now appears that the differences between cultured insect and mammalian cells (Agathos, 1991) are not as overwhelming as it was thought even a few years ago. For example, the myths of insect cells' excessive fragility to hydrodynamic shear and exceptionally high oxygen demand are not supported by experimental data (Maiorella et al., 1988). In fact, it is becoming progressively clearer that resistance to shear is, to a very large extent, a function of the identity of the insect cell line and of the presence of media ingredients, just as is true for other animal cells (Goldblum et al., 1990; Papoutsakis, 1991).

For insect cell cultivation, as for the in vitro propagation of other animal cells, the two most basic criteria that will continue to be used for improved bioreactor design and scale-up are: (1) the aerobic metabolism of these cells and the physical, mechanical and geometric constraints imposed on the devices used for homogeneous mixing and oxygen supply to the culture, and (2) the desirability of high productivity per unit bioreactor volume, which, ideally, implies both higher rates of product formation per cell and high cell density per unit reactor volume.

In addressing the first set of design considerations, i.e., the satisfaction of the cellular oxygen demand during both the growth and the infection stage, when oxygen uptake rates are higher (Schopf et al., 1990; Kamen et al., 1991), the means of aeration should be compatible with the relative sensitivity to shear expected of the insect cells in comparison with microbial cells, based on the insect cells' larger size and lack of cell wall. Additionally, the relative sensitivity to shear among alternative insect cell lines will be an important priority for investigation, since it may dictate different strategies for oxygenation of different cell lines. For instance, the higher fragility of *T. ni* cells compared to *S. frugiperda* cells, which may be due to their larger size and more spindle-like shape, has been independently confirmed by spinner growth experiments (Ogonah et al., 1991), by viscometric experiments (Goldblum et al., 1990), and by experiments aimed at adapting static cell cultures to shake flask cultures (Wu et al., 1990).

A wide variety of agitated reactors with sparging or surface aeration, of airlift reactors, and of bubble columns, spanning a broad range of working volumes (from less than 1 liter to over 40 liters) have already been documented in insect cell cultivation, as seen in several recent reviews (Wu et al., 1989; Agathos et al., 1990; Shuler et al.,

1990; Agathos, 1991; Goosen, 1991). The loss of insect cell viability in agitated and aerated reactors or bubble columns has been primarily attributed to shear forces exerted as a result of air sparging, which can cause cell death either in the area of bubble disengagement in the free liquid surface or in the area of air injection (Tramper et al., 1986; Tramper et al., 1988; Murhammer and Goochee, 1990b). The supplementation of culture media with pluronic polyols (nonionic surfactants) has been shown to be effective in preventing or greatly diminishing insect cell death in agitated and aerated bioreactors (Murhammer and Goochee, 1988;1990b). Although the mechanism of cell damage by hydrodynamic shear is not yet fully understood, it is becoming progressively clear that, in most well-documented cases, cell death actually occurs solely as a consequence of shear forces generated by film drainage around bubbles or by breakup during bubble disengagement at the culture surface (Handa-Corrigan et al., 1989). Repeated failure of several research workers to cultivate insect cells in unsupplemented airlift reactors (Koike and Sato, 1988; Wudtke and Schügerl, 1987) has also implicated shear forces stemming from bubble disengagement. Controlling the diameter of air bubbles to avoid bubble coalescence, and adding pluronic surfactant to the serum-free medium has enabled Maiorella et al. (1988) to successfully cultivate Sf-9 cells in a 21-liter airlift reactor.

For agitated bioreactors, Kunas and Papoutsakis (1990) have attempted to separate the detrimental effects of intense agitation on suspension cultured hybridoma cells from those of air sparging. These workers have shown that two fluid mechanical mechanisms are involved in cell injury and death: the first is present only when there is a gas phase, and is due to bubble breakup either because of vortex formation and gas entrainment or because of direct sparging, whereas the second mechanism causes cell injury in the absence of a gas phase (and, therefore, in the absence of bubbles) only at very high agitation rates (e.g., above 700 rpm in a 2 liter bioreactor with a 7 cm diameter impeller) by stresses in the bulk turbulent liquid. With the latter mechanism, cell damage correlates with Kolmogorov eddy sizes of the same magnitude or smaller than the cell size (9–15 μm). Since, in most agitated bioreactors of practical use for cell cultivation, agitation rates do not normally exceed 300–500 rpm, Papoutsakis (1991) concludes that bubble breakup is the main mechanism of cell damage, as is true for bubble column and airlift bioreactors. It is reasonable to expect that the same mechanisms are responsible for the adverse effects seen in the cultivation of suspended insect cells in agitated and/or aerated bioreactors.

Although there are some indications that the protective effect of pluronic surfactants correlates well with the hydrophilic-lipophilic balance of these block co-polymers and is perhaps acting by some kind of interaction with the cell membrane (Murhammer and Goochee, 1990a), the preponderance of evidence lies mostly on the side of a physical mechanism of protection (Handa-Corrigan, 1989; Michaels et al., 1991). The physical mechanism appears to imply that the additives change the average bubble size and affect the draining and bursting properties of the bubbles at the point of disengagement. In a thoughtful recent review, Papoutsakis (1991) has analyzed the available evidence for mechanisms of shear-induced cell death and the protective effect of media additives, and has shown that, in many cases, both fluid mechanical mechanisms and biological mechanisms (fast acting cell membrane interactions and slower acting metabolic effects) may mediate this apparent protection. Elegant current experimentation aimed at visualizing the processes associated with bubble bursting (e.g., film drainage, liquid jet ejection) through video microscopy and fluorescent dye activation (J. Chalmers; D. I. C. Wang, personal communications) promises to add significantly to our knowledge of the phenomenon of cultured animal cell death under realistic bioreactor conditions, including, perhaps, the eventual elucidation of the protective role played by media additives such as pluronic polyols, carboxymethylcelluloses, serum proteins, etc.

Strategies aimed at outright avoiding of direct sparging or, more generally, of any extended liquid-gas interface, such as attempts to aerate the insect cell culture through semipermeable silicon or teflon tubing (Eberhardt and Schügerl, 1987; Büntemeyer et al., 1987) are of limited practical use for scale-up. On the contrary, with better understanding of the physical and biological basis of cell death due to shear, standard agitated and airlift reactors can be utilized in the future, possibly with small modifications, for large-scale insect cell culture. The retrofitting of such microbial cultivation vessels with low-shear agitators (marine propellers or hydrofoils), the introduction of simple ring spargers providing noncoalescing bubbles, the strategic placement of spargers with respect to the agitators, the complementation between efficient surface aeration and nonaggressive sparging, and the reasoned use of shear protective media additives, should make the scale-up of insect cell culture quite feasible, without having to resort to specialized bioreactor designs.

The second set of considerations in insect cell bioreactor development, namely the attainment of high volumetric productivity, will certainly follow along the general lines of current developments in animal cell bioreactors. This goal can be approached not only by the judicious selection of the most productive insect host cell-vector system on a per cell (or, as seen above, on a per unit cell protein) basis, but also by establishing proven biochemical engineering methodologies, such as high density suspension culture by cell immobilization, perfusion culture, continuous culture with recycle, or combinations of these concepts. For example, *S. frugiperda* cells could be successfully propagated in a 8 liter spin filter bioreactor through continuous perfusion of medium and removal of waste by-products, leading to high cell densities (Weiss et al., 1989). A variation of this technique, in which continuous removal of the spent medium is effected by a hollow fiber filtration module, proved beneficial for high cell density and enhancement of polyhedra formation after infection with a wild type AcNPV (Klöppinger et al., 1990). In implementing high density suspension culture of insect cells, the question will arise whether the actual foreign protein productivity will increase with high cell concentration. There are reports that infection of high density *T. ni* cell culture with AcNPV was inhibited, compared to low density infection (Wood et al., 1982), and that the per cell specific productivity of a recombinant rotaviral capsid protein in Sf-9 cells at high cell densities dropped, compared to that obtained at lower densities (Caron et al., 1990). Typically, low cell densities mean approximately one half to one million cells per milliliter of culture, whereas high cell densities imply typically more than two million cells per milliliter. It is unclear at this stage whether this apparent inhibition is real or artifactual, and, if real, whether the mechanism is associated with depletion of nutrients, including oxygen, accumulation of metabolic inhibitors, or cell to cell contact preventing viral receptor binding. The timely elucidation of these points will lead to optimized bioreaction cultivation and infection protocols.

Cell density increases are also possible via immobilization or attachment to inert matrixes. Abundant production of recombinant β-galactosidase has been shown in high density cultivation of an anchorage dependent *T. ni* strain on glass beads in a packed bed bioreactor (Shuler et al., 1990). Also, Sf-9 cells have attained very high cell densities and baculovirus titers per unit volume of support matrix when immobilized in alginate-polylysine microcapsules (King et al., 1988). Ultimately, the commercial attractiveness of these innovations will rest squarely on the actual densities of viable cells attained per unit bioreactor volume, on the translation of these values to correspondingly high productivities of recombinant protein or of baculovirus titer, and on the ease of product recovery and purification, compared to simple suspension culture.

The mode of bioreactor operation will also influence the overall productivity of the process. Continuous insect cell cultivation has been proposed in order to attain high and uniform productivity. In this process, two (Kompier et al., 1988) or three (van Lier et al., 1990) bioreactors are run in series, the first for growth of uninfected Sf-9 cells, followed by one or two bioreactors for cell infection and virus production. By all measures (polyhedra per cell, polyhedra-containing cells, or nonoccluded virus), virus productivity has been shown to drop with time, thus limiting the useful period of continuous operation to approximately one month (Tramper et al., 1990). The reason for this situation of diminishing returns has been recently attributed to the so called passage effect, i.e., the well-known reduction of AcNPV infectivity upon prolonged passage through multiple infection cycles (Tramper et al., 1990). It has been recently shown that the basis for this effect is the progressive enrichment of the serially passaged virus in defective interfering particles, i.e., deletion mutant particles interfering with the production of fully active AcNPV (Kool et al., 1990; Wickham et al., 1991). Thus, low passage viral inocula will need to be used in future developments of commercial continuous or semicontinuous bioprocessing schemes. On the other hand, the recently published nonlytic system of stably transformed lepidopteran cells in culture (Jarvis et al., 1990) and the also newly developed system of stably transformed *Drosophila* cells in culture (Culp et al., 1991) lend themselves naturally to continuous bioreactor propagation, since the principle behind these systems is totally different from the lytic BEV-based system: the foreign gene is inserted in the insect cell chromosome, and, in the *Drosophila* system, it is highly amplified (Culp et al., 1991). Because of this feature, high rate perfusion culture of *Drosophila* cells will be an important tool in increased recombinant productivity, given the growth-associated production pattern of the recombinant product (Schütz et al., 1991).

9.4 DEVELOPMENT OF CULTURE MEDIA

Further progress in the generalized use of insect cell culture for commercial applications will depend on the development of cheap, conveniently formulated culture media. The desirability of completely defined, serum-free and low protein or protein-free media has been amply recognized, and is one of the clearest trends in insect cell technology today (Vaughn and Fan, 1989; Vaughn and Weiss, 1991). This trend will continue not only for the purpose of eliminating serum, which is expensive, of variable composition, unnecessary for insect cell growth, and often interferes with downstream processing, but also in order to improve the nutrition of the cell population and its susceptibility to infection, and to supply ingredients necessary for surviving the rigors of the bioreactor environment, as seen above.

Careful studies addressing the detailed nutritional and biosynthetic requirements of insect cells in culture, as well as the development of adaptation protocols, will be necessary for serum-free medium design and optimization, especially in the area of commercial serum replacements and shear protective agents (Vaughn and Fan, 1989; Hink, 1991). An additional direction, especially for the mass production of commodity products like biopesticide baculoviruses, will be the design of universal or totipotent media that can be easily formulated, autoclavable, and can support the growth of several insect cell lines (Koike and Sato, 1988).

The selection of serum-free media, including the basal medium formulation or the partial elimination of serum, will often be influenced by the level of foreign protein production and by the degree or type of post-translational processing, which might be compromised. For instance, there is evidence that high serum concentration in serum-supplemented TNM-FH medium (e.g., 10%), while beneficial for fast cell growth, may not be as conducive to high recombinant β-galactosidase expression in AcNPV-infected Sf-9 cells as low levels of serum (e.g., 0.5%), as shown by Broussard and Summers (1989). Conversely, less than 4% serum in the same medium adversely affected the production of recombinant erythropoietin. The best medium formulation for production of this protein was TNM-FH medium containing 4% serum and only one fifth of the normal concentrations of yeastolate and lactalbumin hydrolysate (Quelle et al., 1989). Also, the complete absence of serum brought about a decrease in the yield of a recombinant rotaviral protein, even though the IPL-41 serum-free medium used supported a high Sf-9 cell growth rate as shown by Caron et al. (1990). These workers also noted an increased sensitivity of the cells to shear-induced injury in serum-free medium compared to serum-supplemented media, as expected from the absence of protective additives (serum proteins). Thus, in most cases, serum-free media will require a well-calibrated complement of cell protective additives (e.g., pluronic polyols) for use in agitated or aerated vessels. Additionally, careful studies will need to be conducted in addressing the oxygen requirements of the cell line in serum-free versus serum-supplemented media, as there is preliminary evidence indicating considerably higher oxygen uptake rates for Sf-9 cells in serum-free than in serum-containing media (W. Hensler, V. Singh, and S. N. Agathos, unpublished observations). Finally, in the current absence of adequate data, the influence of media formulation on the form and extent of posttranslational modification (e.g., glycosylation, folding, secretion, etc.) of relatively complex recombinant proteins will have to be addressed in the immediate future.

CONCLUDING REMARKS

The previous sections addressed a number of fronts where new knowledge will continue being generated in the area of insect cell-based biotechnology. From all available indications it is safe to assume that insect cell culture engineering is here to stay, and that it promises to expand our horizons in both fundamental understanding of BEV-driven gene expression and in efficient production of valuable products. In only a few years, foreign gene expression in cultured insect cells has emerged as a major methodological approach in the arsenal of today's biotechnologist, and, if current research and development productivity in academia and in the private sector is any indicator, the trend will continue toward further expansion of this technology for commercially important applications.

The design of specialized as well as all-purpose baculovirus vectors will offer even more power to high level gene expression in insect cells. The choice of cell lines will be influenced by the level and type of protein expression, including posttranslational modifications. Even though our knowledge of this area is still in its infancy, it is known that insect cells glycosylate proteins somewhat differently from mammalian cells (Kuroda et al., 1990; Davidson et al., 1990) and there are indications that there are differences in the pattern of posttranslational processing of the same protein by different insect cell lines (Hink et al., 1991). This type of research will influence to an important degree the future of insect cell culture engineering, since it is conceivable that, at least in some therapeutic or diagnostic applications, the absolute structural identity between insect cell-derived and human cell-derived protein may make a decisive difference in favor of the latter. Nonetheless, the preponderance of the currently available evidence points to the fact that the overwhelming advantage of the insect cell-BEV system lies in its very abundant expression level of biologicals which are *functionally* identical to their native form (Luckow, 1991). Hence, further development and commercialization of this tool for industrial applications, including appropriate bioreactor scale-up, will be amply justified, provided, of course, that the relevant regulatory agencies are cooperative in this area.

Despite the obvious attraction of therapeutic or other medically oriented proteins as possible high value products targeted for expression in insect cell culture, we should not overlook the reemerging potential of this technology for the production of genetically engineered pest control agents (Bishop, 1989; Wood and Granados, 1991). Exciting innovations in this area include, among others, the development of a recombinant *egt* deletion mutant AcNPV with superior pesticide capacity against infected larvae (O'Reilly and Miller, 1991), and the construction of a much more lethal AcNPV against lepidopteran larvae by incorporating toxin-producing genes from a mite (Tomalski and Miller, 1991) or a scorpion (Stewart et al., 1991) into the viral genome. Indeed, with a progressively favorable regulatory climate and in view of encouraging preliminary field test results (Wood and Granados, 1991), the use of massive quantities of viral insecticides as an environmentally desirable alternative to toxic synthetic pesticides may yet bring insect cell culture engineering to the forefront of the new industrial biotechnology revolution, and transform it into a dominant industrial practice for the next century.

REFERENCES

Agathos, S. N., Jeong, Y-H., and Venkat, K. (1990). *Ann. N.Y. Acad.Sci.,* *589*:372–398.

Agathos, S. N. (1991). *Biotechnol. Adv., 9*:51–68.

Betenbaugh, M. J., Balog, L., and Lee, P-S. (1991). *Biotechnol. Prog., 7*:462–467.

Bishop, D. H. L. (1989a). *Trends Biotechnol., 7*:66–70.

Bishop, D. H. L. (1989b). *Pest. Sci., 27*:173–189.

Broussard, D. A. and Summers, M. D. (1989). *J. Invertebr. Pathol., 54*:144–150.

Büntemeyer, H., Bodeker, B. G. D., and Lehmann, J. (1987). "Modern Approaches to Animal Cell Technology," Proceedings of the 8th Meeting of the European Society for Animal Cell Technology, Tiberias, Israel, 1987, (R. E. Spier and J. B. Griffiths, eds.), Butterworths, Kent, U.K. pp. 411–419.

Caron, A. W., Archambault, J., and Massie, B. (1990). *Biotechnol. Bioeng., 36*:1133–1140.

Culp, J. S., Johansen, H., Hellmig, B., Beck, J., Matthews, T. J., Delers, A., and Rosenberg, M. (1991). *Bio/Technology, 9*:173–177.

Davidson, D. J., Fraser, M. J., and Castellino, F. J. (1990). *Biochemistry, 29*:5584–5590.

Eberhardt, U. and Schügerl, K. (1987). *Devel. Biol. Standard., 66*:325–330.

Fraser, M. J. (1989). *In Vitro Cell. Devel. Biol., 25:*225–235.

French, T. J. and Roy, P. (1990). *J. Virol., 64:*1530–1536.

Goldblum, S., Bae, Y-K., Hink, W. F., and Chalmers, J. (1990). *Biotechnol. Prog., 6:*383–390.

Goosen, M. F. A. (1991). *Can. J. Chem. Eng., 69:*450–456.

Granados, R. R. and Hashimoto, Y. (1989). *Invertebrate Cell System Applications,* Vol. 2 (J. Mitsuhashi, ed.), CRC Press, Boca Raton, Florida, pp. 3–13.

Griffiths, B. (1990). *Cytotechnology, 3:*109–116.

Handa-Corrigan, A., Emery, A. N., and Spier, R. E. (1989). *Enzyme Microb. Technol., 11:*230–235.

Hink, W. F. and Hall, R. L. (1989). *Invertebrate Cell System Applications,* Vol. 2, (J. Mitsuhashi, ed.), CRC Press, Boca Raton, Florida, pp. 233–269.

Hink, W. F. (1991). *In Vitro Cell. Develop. Biol., 27A:*397–401.

Hink, W. F., Thomsen, D. R., Davidson, D. J., Meyer, A. L., and Castellino, F. J. (1991). *Biotechnol. Prog., 7:*9–14.

Hu, W-S. and Peshwa, M. V. (1991). *Can. J. Chem. Eng., 69:*409–420.

Iatrou, K., Meidinger, R. G. and Goldsmith, R. (1989). *Proc. Natl. Acad. Sci. U.S.A., 86:*9129–9133.

Jarvis, D. L., Fleming, J. G. W., Kovacs, G. R., Summers, M. D., and Guarino, L. A. (1990). *Bio/Technology, 8:*950–955.

Kamen, A. A., Tom, R. L., Caron, A. W., Chavarie, C., Massie, B., and Archambault, J. (1991). *Biotechnol. Bioeng., 38:*619–628.

King, G. A., Daugulis, A. J., Faulkner, P., and Goosen, M. F. A. (1988). *Biotechnol. Lett., 10:*683–688.

Klöppinger, M., Fertig, G., Fraune, E., and Miltenburger, H. G. (1990). *Cytotechnology, 4:*271–278.

Koike, M. and Sato, K. (1988). *Invertebrate and Fish Tissue Culture* (Y. Kuroda, E. Kurstak, and K. Maramorosch, eds.), Springer-Verlag, New York, pp. 7–10.

Kompier, R., Tramper, J., and Vlak, J. M. (1988). *Biotechnol. Lett., 10:*849–854.

Kool, M., van Lier, F. L. J., Vlak, J. M., and Tramper, J. (1990). *Agricultural Biotechnology in Focus in the Netherlands,* (J. J. Dekkers, H. C. van der Plas, and D. H. Vuijk, eds.), Pudoc, Wageningen, The Netherlands, pp. 64–69.

Kozak, M. 1990. *Proc. Natl. Acad. Sci. U.S.A., 87:*8301–8305.

Kunas, K. T. and Papoutsakis, E. T. (1990). *Biotechnol. Bioeng., 36:*476–483.

Kuroda, K., Geyer, H., Geyer, R., Doerfler, W., and Klenk, H.-D. (1990). *Virology, 174:*418–429.

Licari, P. and Bailey, J. E. (1991). *Biotechnol. Bioeng., 37:*238–246.

Luckow, V. A. and Summers, M. D. (1988). *Bio/Technology, 6:*48–55.

Luckow, V. A. and Summers, M. D. (1989). *Virology, 170:*31–39.

Luckow, V. A. (1991). *Recombinant DNA Technology and Applications* (A. Prokop, R. K. Bajpai, and C. S. Ho, eds.), McGraw Hill, New York, pp. 97–152.

Lynn, D. E. and Hink, W. F. (1980). *J. Invertebr. Pathol., 35:*234–240.

Maeda, S. (1989). *Annu. Rev. Entomol.*, *34*:351–372.

Maiorella, B., Inlow, D., Shauger, A., and Harano, D. (1988). *Bio/Technology*, *6*:1406–1410.

Matsuura, Y., Possee, R. D., and Bishop, D. H. L. (1986). *J. Gen. Virol.*, *67*:1515–1529.

McIntosh, A. H. and Ignoffo, C. M. (1989). *J. Invertebr. Pathol.*, *54*:97–104.

Michaels, J. D., Petersen, J. F., McIntire, L. V., and Papoutsakis, E. T. (1991). *Biotechnol. Bioeng.*, *38*:169–180.

Miller, L. K. (1988). *Annu. Rev. Microbiol.*, *42*:177–199.

Murhammer, D. W. and Goochee, C. F. (1988). *Bio/Technology*, *6*:1411–1418.

Murhammer, D. W. and Goochee, C. F. (1990a). *Biotechnol. Prog.*, *6*:142–148.

Murhammer, D. W. and Goochee, C. F. (1990b). *Biotechnol. Prog.*, *6*:391–397.

Ogonah, O., Shuler, M. L., and Granados, R. R. (1991). *Biotechnol. Lett.*, *13*:265–270.

Ooi, B. G., Rankin, C., and Miller, L. K. (1989). *J. Mol. Biol.*, *210*:721–736.

O'Reilly, D. R. and Miller, L. K. (1991). *Bio/Technology*, *9*:1086–1089.

Page, M. J., Hall, A., Rhodes, S., Skinner, R. H., Murphy, V., Sydenham, M., and Lowe, P. N. (1989). *J. Biol. Chem.*, *264*:19147–19154.

Papoutsakis, E. T. (1991). *Trends Biotechnol.*, *9*:316–324.

Persson, B., Kierulff, J., and Emborg, C. (1991). "Production of Biologicals from Animal Cells in Culture" Proceedings of the 10th Meeting of the European Society for Animal Cell Technology, Avignon, France, (R. E. Spier, J. B. Griffiths, and B. Meignier, eds.), Butterworth-Heinemann, Oxford/Boston, pp. 381–384.

Quelle, F. W., Caslake, L. F., Burkert, R. E., and Wojchowski, D. M. (1989). *Blood*, *74*:652–657.

Schopf, B., Howaldt, M. W., and Bailey, J. E. (1990). *J. Biotechnol.*, *15*:169–186.

Schütz, C., Jäger, V., Driesel, A. J., and Wagner, R. (1991). "Production of Biologicals from Animal Cells in Culture" Proceedings of the 10th Meeting of the European Society for Animal Cell Technology, Avignon, France, 1990, (R. E. Spier, J. B. Griffiths, and B. Meignier, eds.), Butterworth-Heinemann, Oxford/Boston, pp. 460–465.

Shuler, M. L., Cho, T., Wickham, T., Ogonah, O., Kool, M., Hammer, D. A., Granados, R. R., and Wood, H. A. (1990). *Ann. N.Y. Acad. Sci.*, *589*:399–422.

Stewart, L. M. D., Hirst, M., López Ferber, M., Merryweather, A. T., Cayley, P. J., and Possee, R. D. (1991). *Nature*, *352*:85–88.

Takehara, K., Ireland, D., and Bishop, D. H. L. (1988). *J. Gen. Virol.*, *69*:2763–2778.

Tomalski, M. D. and Miller, L. K. (1991). *Nature*, *352*:82–85.

Tramper, J., Williams, J. B., Joustra, D., and Vlak, J. M. (1986). *Enz.Microb. Technol.*, *8*:33–36.

Tramper, J., Smit, D., Straatman, J., and Vlak, J. M. (1988). *Bioproc. Eng.*, *3*:37–41.

Tramper, J., van den End, E. J., de Gooijer, C. D., Kompier, R., van Lier, F. L. J., Usmany, M., and Vlak, J. M. (1990). *Ann. N.Y. Acad. Sci., 589:*423–430.

van Lier, F. L. J., van den End, E. J., de Gooijer, C. D., Vlak, J. M., and Tramper, J. (1990). *Appl. Microbiol. Biotechnol., 33:*43–47.

Vaughn, J. L. and Fan, F. (1989). *In Vitro Cell. Develop. Biol., 25:*143–145.

Vaughn, J. L. and Weiss, S. A. (1991). *BioPharm, 4:*16–19.

Volkman, L. E. and Knudsen, D. L. (1986). *The Biology of Baculoviruses,* Vol. 1, (R. R. Granados and B. A. Federici, eds.), CRC Press, Boca Raton, Florida, pp. 109–127.

Weiss, S. A., Belisle, B. W., DeGiovanni, A., Godwin, G., Kohler, J., and Summers, M. D. (1989). *Continuous Cell Lines as Substrates for Biologicals, 70:*271–279.

Weyer, U. and Possee, R. D. (1989). *J. Gen. Virol., 70:*203–298.

Wickham, T. J., Davis, T., Granados, R. R., Hammer, D. A., Shuler, M. L., and Wood, H. A. (1991). *Biotechnol. Lett., 13:*483–488.

Williams, G. V., Rohel, D. Z., Kuzio, J., and Faulkner, P. (1989). *J. Gen. Virol., 70:*187–202.

Wood, H. A., Johnston, L. B. and Burand, J. P. (1982). *Virology, 119:*245–254.

Wood, H. A. and Granados, R. R. (1991). *Annu. Rev. Microbiol., 45:*69–87.

Wu, J. Y., King, G., Daugulis, A. J., Faulkner, P., Bone, D. H., and Goosen, M. F. A. (1989). *Appl. Microbiol. Biotechnol., 32:*249–255.

Wu, J., King, G., Daugulis, A. J., Faulkner, P., Bone, D. H., and Goosen, M. F. A. (1990). *J. Ferment. Bioeng., 70:*1–5.

Wudtke, M. and Schügerl, K. (1987). "Modern Approaches to Animal Cell Technology," Proceedings of the 8th Meeting of the European Society for Animal Cell Technology, Tiberias, Israel, (R. E. Spier, and J. B. Griffiths, eds.), Butterworths, Kent, U.K., pp. 297–315.

Index